THE MORAL WAGER

Philosophical Studies Series

VOLUME 108

The titles published in this series are listed at the end of this volume.

THE MORAL WAGER

Evolution and Contract

MALCOLM MURRAY
University of Prince Edward Island, Charlottetown, Canada

A C.I.P. Catalogue record for this book is available from the Library of Congress.

ISBN 978-1-4020-5854-7 (HB)
ISBN 978-1-4020-5855-4 (e-book)

Published by Springer,
P.O. Box 17, 3300 AA Dordrecht, The Netherlands.

www.springer.com

Printed on acid-free paper

PREFACE

In the following chapters, I offer an evolutionary account of morality and from that extrapolate a version of contractarianism I call *consent theory*. Game theory helps to highlight the evolution of morality as a resolution of interpersonal conflicts under strategic negotiation. It is this emphasis on strategic negotiation that underwrites the idea of consent. Consent theory differs from other contractarian models by abandoning reliance on rational self-interest in favour of evolutionary adaptation. From this, more emphasis will be placed on consent as natural convergence rather than consent as an idealization. My picture of contractarianism, then, ends up looking more like the relativist model offered by Harman, rather than the rational (or pseudo-rational) model offered by Gauthier, let alone the Kantian brands of Rawls or Scanlon. So at least some of my discussion will dwell on why it is no loss to abandon hope for the universal, categorical morality that rational models promise.

In the introduction, I offer the betting analogy that underwrites the remaining picture. There are some bets where the expected utility is positive, though the odds of winning on this particular occasion are exceedingly low. In such cases, we cannot hope to give an argument that taking the bet is rational. The only thing we can say is that those predisposed to take this kind of bet on these kinds of occasions will do better than those with other dispositions, so long as such games occur often enough. The lure of morality is similar. Moral constraint is a bad bet taken in and of itself, but a good bet when examined statistically. The game of morality occurs whenever strategic negotiation takes place, and since this occurs often enough for social creatures such as us, an attraction for moral dispositions exists.

Components of this work have appeared in print before. Sections 3.1 to 3.3 are modified from my "A Catalogue of Mistaken Interests: Reflections on the Desired and the Desirable," *International Journal of Philosophical Studies* 11/1 (2003): 1–23. Chapter 7 is a modified and expanded version of my "Concerned Parties: When Lack of Consent is Irrelevant," *Public Affairs Quarterly* 18/2 (2004): 125–40.

While writing *The Moral Wager*, I have greatly benefited from discussions with Jason Alexander, Richmond Campbell, David Chan, David Copp, Peter Danielson, Daniel Dennett, Cynthia Dennis, Susan Dimock, Casey Dorrell, Stephen Finlay, David Gauthier, Louis Groarke, Carl Hahn, Marcus Jardine, Esther Kemp, Chris Maddocks, Thaddeus Metz, Jan Narveson, Toni Rønnow-Rasmussen, Chris Tucker, Peter Vallentyne, David Velleman, Paul Viminitz, and Sheldon Wein. This is not to say they agree with me. Thanks also to Stephen Hetherington, Series Editor at Springer, who has helped nurse this work to its present state. Most of all, however, I thank Pat and Emma Murray, who appear to love me despite my moral philosophy.

CONTENTS

INTRODUCTION

The Gist

Few believe that morality is a bad thing, but there is no consensus on what makes it a good thing. I offer a picture of morality that eschews rational justification in favour of evolutionary dynamics. The difficulty with naturalized approaches is in deriving normative principles from non-normative beginnings. Evolutionary forces cannot dictate normative duties. I argue that a principle of consent, what I call *consent theory*, is the best semantic representation of the forces of replicator dynamics. By "best semantic representation," I mean that, should we try to form a moral principle that best represents the dynamics we see going on in evolutionary models, something akin to, "Be a conditional cooperator!" is it. From this, at least two things need to be addressed: (1) Offering the best semantic representation is short of offering a "justification" of the normative principle of consent. The term "justification," however, is more fitting within the language that rational appraisal monopolizes, but evolutionary success does not mimic the rational solution. (2) The principle of consent is shy of the kinds of moral duties commonly associated with social norms. Evolutionary ethics need not have a one-to-one correlation with currently held norms, however. Some accepted norms may be overextended heuristics. So I shall argue.

By analyzing the evolutionary dynamics of interpersonal relations in a variety of games, we discover that strategies reaching equilibria have something in common: such strategies are prone to conditionally cooperate with other cooperators. Moral advice, then, becomes "imitate these strategies in your dealings." Successful players in evolutionary games operate *as if* following the normative advice: *Don't do unto others without their consent.* More fully: *Don't do unto others without their consent, or expected consent, or what they would consent to if they were capable of giving consent, so long as these others abide by this same norm.* The evolutionary success of strategies that operate as if following this advice is what gives it its normative force.

If the principle of consent is the best semantic representation of evolutionary dynamics, a number of implications to moral theory follow. First, we cannot view morality as an unconditional strategy. Moral theory cannot tell us what to do independently of what other agents are doing. This will make it difficult to view morality as a categorical imperative. Hypothetical imperatives bespeak of conditional strategies, categorical imperatives do not. Replicator dynamics do not favour unconditional strategies. Secondly, we cannot hope to offer rational justification for adopting the moral advice. Evolutionary dynamics track successful strategies over time and across generations, whereas rationality demands utility maximization in the current situation.

The Bet Analogy

Generally people who are moral tend to do better than people who are not. Saying "generally" here is an admission that being moral is not demonstrably rational in all cases. This is where Glaucon and Adeimantus went astray: they wanted a proof of the benefits of morality in every single case. In fact, we could even admit that being moral is *never* rational in any given case without undermining the statistical claim. Such an admission may initially strike one as untenable. After all, it is incoherent to claim, "Although no swans are white, don't worry, statistically swans are white" – assuming the existence of swans. But consider a parallel case: the average number of children may well be 2.5; this is not refuted by noting that no family has 2.5 children. Or consider the claim, "A woman gives birth every seven seconds." Here is a claim that is false in any individual case, yet true in the statistical case. That is, it is justified as a statistical claim, yet not justifiable as a claim about a particular woman. Something similar is going on with moral advice: it cannot be justified in the particular instance, but can be justified statistically.

Analogously, there are some bets where the expected utility is positive, even though the odds of winning on this particular occasion are exceedingly low. In such cases, we cannot hope to give an argument that it is rational to take the bet on this occasion. The only thing we can say is that those predisposed to take this kind of bet on these kinds of occasions will do better than those with other dispositions, so long as such games occur often enough. The lure of morality is similar. Moral constraint is a bad bet taken in and of itself, but a good bet when examined statistically. Consider, for example, the following bet.

> Pick one card from a full, normal deck of cards. If it is a seven, you get \$100, if it is a spade, you get \$25. Otherwise you pay \$10.

There are two ways of thinking about this. To maximize your expected utility, you should take the bet. That is, statistically, if you played this game a sufficient number of times, you would stand to win in the long haul. On average, you would stand to gain \$7.04 each time you played the game [(\$100 × 4/52) + (\$25 × 13/52) – (\$10 × 36/52)]. So in one sense, you should take the bet. But if you think that you will thereby make \$7.04 by taking the bet on this occasion, you have reasoned amiss. In fact, your chances of losing \$10 playing this particular game on this particular occasion is 69 % (36/52). Taking a bet where you have a 69 % chance of losing \$10 when \$10 matters to you is not rationally justifiable. So even though expected utility theory tells you to take the bet, straightforward rationality dictates you avoid it. Nevertheless, if the game in which the bet is offered is open ended, those who take the bet stand to gain in the long haul over those who do not. It is that process evolutionary forces track. Similarly, despite rationality's speaking against being moral, evolutionary fitness speaks in morality's favour.

Please do not misconstrue me. I am not saying those who are inclined to take bets do better than those who are disinclined to take bets. There are many bad bets. Rather, those who are inclined to take bets when the expected utility is positive will tend to do better than those who take bets indiscriminately and those who do not take bets on principle. Successful agents will be conditional bettors, not bettors *simpliciter*. The evolutionary benefits of morality are similar: conditionally cooperative agents will do better than both unconditionally cooperative agents and unconditionally uncooperative agents – under suitable conditions.

An affinity to Hume lurks here. Hume said that reason cannot back our conviction that the sun will come up tomorrow but, fortunately for us, nature foists the belief upon us anyway. Something similar – with an important qualification – may be said about our moral beliefs. Moral actions cannot rationally be justified, but fortunately for us, certain moral beliefs are thrust upon us whether we like it or not. There are differences between the two cases, to be sure. We cannot help but believe in causation, induction, and in the existence of external objects independent of our senses, whereas we can avoid belief in morality. Perhaps the odds of social success for the outright immoral are low, but belief in morality is not *forced* the way belief in causation is *forced*. Another, though related, difference concerns the wide uniformity in belief in one's senses across cultures; the uniformity is less wide concerning moral beliefs across cultures. These differences do not matter for my purposes, however. Although neither moral beliefs nor belief in our senses can be rationally justified, they both have evolutionary fit.

This insight into the fortunate nature of morality is not to be taken too far. Once we pay closer attention to the statistical utility of morality, we can begin to tease apart evolutionary benefits from cultural exaggerations. What people mean by morality and what aspect of morality is evolutionarily beneficial are not necessarily the same. In Chapter 2, for example, I argue that although people normally assume morality must be categorical, they are mistaken. In response to the error that people are in about morality, I do not offer an error-theory. Error theorists maintain that people are in error about morality, but that they understand morality the only way one can. For example, one cannot believe in witches without also believing in supernatural powers. To dismiss the existence of supernatural powers is to dismiss the belief in witches. We do not modify witch talk to take into account the non-existence of supernatural powers – we abandon witch talk entirely. But one can be in error about morality in less fundamental ways – even if the view held in error is deemed fundamental by those holding the erroneous view. After all, many theists hold that God is fundamental to morality, but rejecting belief in God does not mean we must reject belief in morality. That those very theists would think so is not informative.

I offer a partial-error theory. Moral categories have use, and forming moral heuristics will also have use, but people's estimations of morality from those heuristics is where they go astray. They have overextended the heuristic. By highlighting the mechanics of how morality has use in terms of evolutionary fit, we highlight all that we are entitled to say about morality.

An assumed implication of naturalized ethics is the collapse of morality into an absolute relativism. If moral discourse cannot be rationally justified, then neither can any moral discourse be rationally criticized. If nothing is right, nothing is wrong. Whatever one's view of morality is, it must define right from wrong, and so any theory that collapses to simple relativism is merely an obfuscated confession of moral nihilism. Although the naturalized, non-rational picture of morality I wish to present is relativistic, it is not the crass relativism that collapses into nihilism. Hume used the word "fortunately" and that word is key. Much of our understanding of morality is mistaken, but not all. The common bits about moral discourse turn on conflict resolution of certain strategic interactions. Although it may not be individually rationally justifiable for agents to adopt moral dispositions, evolution favours agents who adopt a conditional cooperative strategy when faced with conflicts of strategic interaction. Some might be tempted to say that conditional cooperative strategies are justified to the extent that such strategies are evolutionarily stable. But this is not the right way of looking at things, as the bet analogy above demonstrates. The justification cannot be aimed at the individual actor. There is no justification for *her* to be moral when being immoral pays her greater dividends in the particular case. The "justification" if it can be called that, is aimed more at the level of statistical trends. We can explain why morality is prevalent despite its unjustifiability on individual terms. Being moral cannot be rationally justified, but saying this does not commit us to grope for some non-natural property accessed by opaque intuition faculties.

Chapter by Chapter

My endorsement of evolutionary ethics commits me to a form of reductive natural irrealism. In Chapter 1, "Irrealism," I explain what reductive natural irrealism entails. To be a moral realist is to believe that discourse about ethics is a discourse about objective moral properties and facts. One may believe discourse about morality is necessarily discourse about objective moral properties and facts, but that there are no such things (error theorists). Those moral realists who are not error theorists believe these objective moral properties exist in fact. Some of these moral realists believe these moral properties are non-natural (Moore, e.g.), and others believe they are natural (Bloomfield, e.g.). An irrealist denies that moral discourse requires the belief in objective moral properties, and denies the existence of objective moral properties or facts. Some irrealists go further and deny that moral discourse has any propositional content at all (non-cognitivsm). Other irrealists admit that moral discourse may be true or false; they simply deny that truth-value of moral discourse has any connection to objective natural or non-natural properties. Of this latter group, a further distinction may be made, that between reductionists and non-reductionists. Non-reductionists believe that although moral discourse can be true or false, moral discourse is still purely evaluative (Timmons and Gibbard, e.g.). My position is that moral discourse fully reduces to natural descriptive facts of the world. The bits commonly held to be left over are deemed eliminable. Holding on to those bits is where the error lies; not in failing to accommodate them. Such a move permits a form of relativism in moral appraisal. It also precludes speaking of the natural facts and properties as being themselves moral facts or properties. Thus, I carve out a reductive naturalist irrealism in Chapter 1. I also distinguish well-being discourse from moral discourse. In this book, I am only concerned with the latter. Morality, I proclaim, concerns interpersonal relations, not intrapersonal relations.

Key philosophers discussed in this chapter include Simon Blackburn, Paul Bloomfield, David Copp, David Hume, Peter Railton, Mark Timmons, and Russ Shafer-Landau.

In Chapter 2, "Against Moral Categoricity," I highlight how normal moral discourse is flawed. Normal moral discourse holds that morality is fundamentally a categorical imperative. This is wrong on two counts: it is not a categorical imperative, nor is it fundamentally so. That morality need not be fundamentally viewed as categorical distinguishes my position from Richard Joyce's. Joyce holds that once we jettison categoricity, we become error theorists. I argue that a partial-error theory is possible.

Key philosophers discussed in this chapter include Phillipa Foot, Richard Joyce, Immanuel Kant, Christine Korsgaard, J.L. Mackie, Hilary Putnam, and John Searle.

In Chapter 3, "Self-Interest," I examine the implications of understanding morality as a system of hypothetical imperatives. If moral advice is hypothetical, it must be predicated on something. The standard offering is self-interest. Generally critics argue that we cannot draw morality out of self-interest. I argue, instead, that the role of self-interest is largely misconstrued. It is only within the confines of interpersonal interaction where self-interest comes into play. To rely too heavily on self-interest *simpliciter* would mean that morality has more to do with personal well-being, and less to do with a social convention geared to resolving conflicting interests. Morality's role is not to serve preferences, but to solve interpersonal preference conflicts. As a solver of interpersonal preference conflicts, morality must be non-partisan to preferences themselves. (The retort that solving interpersonal preference conflicts is a necessary, but not sufficient, condition of morality will be addressed in Chapter 8.) At the end of Chapter 3, I highlight how evolutionary game theory is well designed to track the success of moral strategies – no matter what one's preferences are.

Key philosophers discussed in this chapter include Richard Brandt, Joseph Butler, Harry Frankfurt, David Gauthier, David Griffin, Thomas Hobbes, Gregory Kavka, Christine Korsgaard, Derek Parfit, Joseph Raz, David Schmidtz, David Wiggins, and Susan Wolf.

In Chapter 4, "Rationality," I argue against justifying morality in terms of rationality. In this chapter, I introduce game theory: that is, I present a few simple tables, and some elementary game theoretic calculations. Specifically, I highlight David Gauthier's argument that conditional cooperation (CC) in Prisoner's Dilemmas (PDs) is more rational than unconditional defection (UD). Importantly, if morality is rational, we can answer Hobbes's Foole in terms the Foole should be able to endorse. I argue, however, that the CC strategy is not rational. Concerning the first, the success of CC depends on the robustness of the conditions under which CC prevails. For CC to prevail, the following conditions must be met.

(1) There are enough other CCs with whom to interact.
(2) There are no computation costs that CCs have that UDs lack.
(3) The ability for CCs to correctly identify UDs is sufficiently accurate.
(4) No other disposition is allowed into the game.

The reply to (1) depends on our accepting the replies to (2), (3), and (4), and the replies to (2), (3), and (4) depends on our accepting the reply to (1). Enough CCs can get into the mix *only if* the computation and detection costs are not penal, or so long as there are enough CCs in the mix, or so long as certain other kinds of dispositions are not in the mix. Assuming even minimal computation and detections costs, those initial CC agents would have to be irrational: they will do worse than being a UD. CC's rationality is predicated on the irrationality of the first CC agents. Beyond this, even within the parameters that Gauthier sets for us, Gauthier requires a broader understanding of rationality than we may be used to. As Ken Binmore notes, if it is rational to cooperate in a PD, that shows merely that it was not a PD. The PD is defined in such a way that it is always irrational to cooperate with another cooperator when unilateral defection will earn the defector greater individual utility. That rational move (defecting against a cooperator) is precisely what CC agents prevent themselves from doing.

Key philosophers discussed in this chapter include Ken Binmore, Peter Danielson, and David Gauthier.

In Chapter 5, "Evolution," I provide an evolutionary picture of how moral dispositions can thrive. Evolutionary models of ethics show how *despite* their irrationality, moral agents are more likely to pass on their genes than non-moral agents. In this chapter, I introduce a few other games, the Ultimatum Game, the game of Chicken, Battle of the Sexes, and a new invention of my own, the Narrow Bridge Game, which may be viewed as a variant on Chicken. Also, I introduce the formula for replicator dynamics. In this sense, it is a slightly more advanced Chapter than the previous, but is still presented simply enough to be understood by non-specialists.

By examining different evolutionary games, we discover a common element in successful strategies. Agents employing rational strategies tend to be those who cannot do well against their own kind. Paradoxically, the more success they have, the worse they will fare. Conversely, successful agents are those who have the ability to cooperate among their own kind (and defect against others). In the short run, admittedly, they do worse compared to strictly rational agents, but given the paradox of success, agents employing conditionally cooperative strategies prevail.

After examining the underlying mechanism of successful strategies in evolutionary games, I defend evolutionary ethics against two sets of problems. The first concerns problems inherent in the modelling parameters. Of the modelling objections, I consider three.

(1) The problem concerning favourable initial conditions that plagued the rationality model has not been avoided by the evolutionary model.
(2) The problem of whether the "winner" is really moral has also not been avoided by the shift to evolutionary models.
(3) We find polymorphisms in reality, but not in game theoretic results.

The second set of problems is more general. Of these, I discuss four.

(1) Advocating irrational strategies seems patently incoherent.
(2) Both rationality and morality are normative notions, whereas evolutionary accounts can only speak in terms of description.
(3) Evolutionary ethics abandons any hope of providing a justification for morality: evolutionary models can offer explanation of moral behaviour only, not justification. Consequently evolutionary ethics cannot motivate the ethically challenged to become moral.
(4) A fourth general worry highlights the contingencies of evolutionary ethics, which returns us, partly, to the worries raised in Chapters 1 and 2.

Key philosophers discussed in this chapter include Robert Aumann, Robert Axlerod, Ken Binmore, Peter Danielson, David Hull, Brian Skyrms, and Elliott Sober. Works by Gerd Gigerenzer and Reinhard Selten also figure prominently.

In Chapter 6, "Consent Theory," I extrapolate a normative ethical principle from examining the common features of successful strategies in evolutionary games. Once we pay closer attention to the statistical utility of morality, we can begin to tease apart evolutionary benefits from cultural exaggerations. What people think they mean by morality and what aspect of morality is evolutionarily beneficial are not necessarily the same. As highlighted in Chapter 5, and following from the rejection of categorical moralities (Chapter 2), the bare normative advice that is consistent with the games examined would be something like: "Be a conditional cooperator!"

In brief, game theory helps to highlight the evolution of morality as a resolution of interpersonal conflicts under strategic negotiation. It is this emphasis on strategic negotiation that underwrites the idea of consent. Consent theory is a brand of contract theory, but I need to distinguish it from contractarian and contractualist theories. There are two basic differences. One follows from our abandoning reliance on rational self-interest in place of evolutionary adaptation. The other difference is that, unlike other contract theories, consent theory emphasizes the primary role of consent. Moral constraint against coerced agreements cannot be derived from any *ex ante* agreement, for the very concept of agreement must presuppose such constraint. To highlight this circularity, I introduce the game of Proposal, which is played prior to entering the prisoner's dilemma. This game is presented as a simple tree, and employs Zermelo's backward induction. Again this is a simplified table geared to the non-specialist. Its point is not to eliminate *ex ante* agreement, but to illustrate how the *ex ante* agreements that contractualists and contractarians speak about already presumes the more basic principle of consent: namely *any act is moral only so long as all concerned, suitably informed, competent agents agree.* (This little principle needs much unpacking to accommodate issues of competency, proxy consent, surprise parties, positive duties, and the like, much of which will be discussed in chapters 7 and 8.)

To emphasize the circularity inherent in consent theory is to expressly admit that the normative advice of consent theory lacks full rational appeal, but such an observation will be uninteresting for those who have accepted the arguments I offer in chapters 1 to 5. The normative principle of conditional consent is not a matter of rational choice. Rather, the concept is a well enmeshed phenotype: it has evolutionary fit. The root of contract theory is *not* that agents consent to moral constraints, but that in order for mutual benefit to arise from strategic negotiations, they must presuppose the binding force of consent.

Key philosophers discussed in this chapter include David Gauthier, Jean Hampton, Gilbert Harman, Anthony de Jasay, Jody Kraus, John Rawls, Joseph Raz, and Thomas Scanlon.

In the next two chapters, I discuss individual problems with consent theory. In Chapter 7, "Concerned Parties," I emphasize the importance of determining who counts as a concerned party. If this is left undefined, consent theory will provide useless advice. My focus in this chapter is aimed at answering a problem that has not received much attention to date. This is because few ethical theories emphasize consent to the degree I recommend. But once we define moral actions according to occurrent consent, we need to be very clear whose consent we are talking about. It is not the case that anyone off the street who whines about an agreement made between people other than herself should be sufficient to render the agreement null or impermissible. The whining must be justified. The dissenter must be affected in the requisite way in order to squelch the agreements of others. The requisite way, typically, is conceived in terms of harm. The appeal to the harm principle is supposed to rule out busybodies who interfere in others' lives for no good reason. That Barney does not like Betty's having a nose ring is not thought to count as warranting the banishing of nose rings. The concept of harm, however, is not clearly defined. Barney may really feel "harmed" by people wearing nose rings, albeit in a non-physical way. It would be simple if consent theorists could speak of only "physical harms" when they speak of "harms," but this is not right. Imposed psychological harms cannot be permissible in moral theory. But since people have different psychological make-ups, what counts as causing psychological harm will vary according to the person and the situation. If someone is harmed by all sorts of things that normal people are not, this individual must be deemed an unreliable measuring rod of harm. Some criterion of normalcy is required, but what should count as "normal harm" is itself too vague and too flexible to be of much help. To distinguish consent that matters from consent that does not matter, I offer the following criterion:

> *Anyone suffering or expected to suffer physical adverse effects is a concerned party. Otherwise, one is a concerned party only so long as the affecter cannot fulfil her desires without use of the affectee's person or property.*

If you satisfy that condition on a particular occasion, what you say matters. If you do not satisfy that condition on a particular occasion, what you say about that occasion does not matter morally.

Key philosophers discussed in the Chapter include: Joel Feinberg, Derek Parfit, Lois Pineau, and Charles Taylor.

In Chapter 8, "Suffering and Indifference," I tackle the problem of altruism. Although consent theory is well positioned to defend negative duties, it seems to lack the ability to defend any positive duties. If two people agree to row a boat, and this action does not adversely affect others not party to the agreement, nothing could be immoral about the act, or so consent theorists would avow. Many strongly protest. It is often supposed that one's moral duty is to give positive aid to the suffering. Two people who consent to row a boat may be immoral, according to these objectors, if a third were drowning and they were rowing the boat away from the victim. Morality, it is claimed, must require more than mere agreements; in fact, morality must impose strictures on the content of particular agreements. If so, consent theory fails to ground morality.

Within the contractarian tradition, "harm" has been defined in relation to a baseline. If an individual is made worse off than she was, this counts as harm. The question before us, then, is whether or not the drowning victim was harmed by the rowers' failure to rescue her. If you are drowning with your wallet in your pocket, your baseline is the state of drowning with your wallet. Thus, failing to save you will not count as harming you since you will still be drowning with your wallet. Taking your wallet before you drown will be "harming" you for it worsens your state relative to your baseline. Death creates complications to this line of thinking. A drowned man's baseline is so low that nothing could worsen his state. Thus, seemingly, taking a dead man's wallet may not be morally inadmissible. Presumably, however, the contents of the wallet go to the dead man's estate, in which case it would be immoral to take the wallet, since it reduces the baseline of any beneficiary. Fine, but accepting that we ought not take the drowning woman's wallet in this case does not solve anything. If we are to read consent theory as a simple principle of non-harm, then allowing the person to drown in this case would be deemed morally permissible by consent theorists. This does not answer the problem; it exacerbates the problem. The complaint is not that consent theory is inconsistent; it is that those who hold it are immoral, or, more formally, that their theory of morality fails to capture morality. Knowing why the two rowers are not technically harming the drowning woman does not convince many that the two rowers are thereby moral. Many hold it is "morally monstrous" that a moral theory will have nothing to say to someone who allows another person to drown. I partly agree, but my agreement is a very qualified sort. First of all, consent theory is not reducible to a mere non-harm principle. This does not solve the problem. By the criterion of a concerned party offered above, the drowning person in our scenario still would not count as a concerned party. Thereby neither her complaint nor the complaints of those standing on shore watching the rowers row away could morally count. What is also needed is a reminder that consent theory is a semantic representation that approximates in terms of a normative principle the evolutionary forces that propel moral dealings. Evolution provides us with broad heuristics aimed at capturing conditional cooperation in diverse situations. The mechanics of heuristics are such that individual agents operate with broad-stroked algorithms only, and some of our deeply held moral convictions lie in the penumbra of these broad strokes. That is, I defend the conventional praise of altruism as part of an extended heuristic. We can admit altruism does not fit the mechanics of replicator dynamics, except that replicator dynamics will favour algorithms based on heuristics, and it is in the broad strokes of these algorithms where altruism gets its foothold. So I shall argue in Chapter 8.

The success of altruism, at any rate, is not demonstrated by the games examined in Chapter 5. Those games demonstrate the success of conditional agents. Altruism appears to be an unconditional strategy. Nevertheless, the norm of altruism may well piggy-back on a successful strategy. In this sense, we can explain the prevailing norm of altruism as an overextended heuristic. To do this, I borrow from the social theories of Boyd and Richerson, who show the successful strategy of coarse-grained imitation. We prefer to adopt successful behaviours, not unsuccessful ones, and some behaviours which offer short-term gain are offset by long-term loss. To avoid this error, imitation of successful agents of previous generations has clear advantages, so long as the environment is not too unstable. Imitation will tend to be broad-stroked, not fine-tuned, and so some traits will be taken on that are either not present in the original model, or not necessary to the model's success. I argue that the social norms of altruism develop through the overextended imitation of successful phenotypes.

This line of argument carries more weight for those whose reputations are more prominent. Prestige members will be more motivated to guard their cooperative reputation, even by exerting behaviours technically unnecessary for conditional cooperation, if merely to thwart misconstrual, and thus defection, by others. The fear of a marred reputation thereby moves the prestigious toward overextension of cooperative behaviour into fuzzy cases of altruism, and the broad-stroked mechanics of mimicry move the plebeians to imitate the overextended altruism of successful members, which in turn coagulates into a norm.

Key philosophers discussed in this chapter include David Brink, Owen Flanagan, David Gauthier, Alasdair MacIntyre, John Rawls, Thomas Scanlon, Peter Singer, and Brian Skyrms. I also rely on studies by Solomon Asch, Robert Boyd and Peter Richerson, Luca Luigi Cavalli-Sforza and Marcus Feldman, J. M. Darley and C.D. Batson, Gerd Gigerenzer and Reinhard Selten, Melvin Lerner, and Konrad Lorenz.

Chapter 1

IRREALISM

1.1. Magic and Morality

Morality, as I shall defend it, depends on conditions that are contingent. These contingent conditions happen to generally apply for most people in most circumstances. This happy coincidence may cloud the contingencies from view. They are there just the same. The contingent nature of morality undermines moral realism. My plan in this chapter is to provide reasons why abandoning moral realism is no loss.

Consider the following analogies. Magic is real so long as we understand magic as a conjurer's trick, not as magic, which is unreal. A sunset is real so long as we are not fooled into believing the sun sets. The sun setting is not real, it is only the appearance of a sun setting given the rotation of the earth upon which the viewer is stationed. But given those factors, the experience of the sun's setting is real. Deja vu is real so long as we do not mean that we have really experienced a time loop. The deja vu experience is real, although we might account for it in terms of misfiring adrenaline. It is true that the Pope has a belief about God's plans so long as we do not misinterpret this as an admission that we believe the Pope's beliefs about God's plans are true. These examples illustrate that one can speak of a reality of an unreal event. Normally such feats of discourse do not trip us up, but they may. For example, someone might complain after our explaining to him how the magic was not really magic, but a conjurer's trick.[1] He might say, "But that isn't magic!" Such a complaint comes from someone who does not follow the shift in meanings of real. To such a person, our response will sound self-contradictory. We will say something like, "Well, it's not magic in *your* sense, but in your sense no magic exists. In our sense, the reality of magic has been perfectly explained." The respondent may reply: "Nonsense, since your sense leaves out my sense of magic, and so your reduction of magic to non-magical events is hardly a successful reduction of magic!" What can we say to such a person? Only "You are being preposterous."

The question about moral realism suffers the same confusion. There is something about morality that is real, but it is nowhere near what moral realists require. It is not even a step in the right direction, since the senses of realism being employed are not only completely different, they are incompatible.

1.2. Moral Epistemology

When people speak of morality, they may mean one of four different things: (1) how people behave, (2) what people say concerning how people ought to behave, (3) how people in fact ought to behave, and (4) what people are entitled to say about how people ought to behave. That is, when people speak about morality, they may mean to speak about (1) moral psychology, (2) moral norms, (3) moral ideals, or (4) moral epistemic entitlement. My main concern is with (4). My concern with (1), (2), and (3) shall be tangential, at best. By the way, I do not mean to imply these divisions are necessarily opposed. It is conceivable that they perfectly overlap. Most people do not murder, most say that murder is wrong, most suppose any moral ideal will include bans against murder, and most suppose we are entitled to speak about bans on murder. But nothing implies that they must be perfectly correlated like that. It is a commonality to point out that what people say they do and what they do often part ways. Others may insist that the baggage of rules around here permit infidelity, but maintain that the moral ideal ought not. Saying this, people distinguish cultural norms from moral ideals. Assuming that no one has a pipeline to the actual, real, objective moral ideal, any moral ideal is a proposed ideal, and any proposal is open to criticism. To criticize a moral ideal by what people in fact do around here, or by what people say they do around here is obviously an inadequate sort of criticism, for any proposal about how things ought to be done around here is compatible with the recognition that that is not how things are done around here. So to criticize a moral ideal, some other sense of morality must be appealed to, and assessing how well this is done is the domain of moral epistemic entitlement. Many ideals may be proffered, but not all of them are sound. To see which are and which are not sound we examine the epistemic process: we seek to discover what we are entitled to say about morality, not merely what we may want to say.

But can there be a *moral* epistemic process? Those who say yes are cognitivists. Those who say no are non-cognitivists. Whether one is a cognitivist or non-cognitivist depends on whether or not one thinks moral statements are truth-evaluatable.[2] If yes, cognitivism. If no, non-cognitivism.

Being a cognitivist does not necessarily commit one to moral realism. Mark Timmons, Allan Gibbard, and Simon Blackburn, for example, are non-descriptivist cognitivists.[3] That is, contra emotivism, moral utterances are truth-functional, but pace emotivism, they are evaluative only, not descriptive. Moral realism entails something more than admitting we can ascribe truth and falsity to moral utterances, it requires that the truth of the moral utterance is independent of our evaluations. Something's being true or false, then, does not necessarily commit one to moral realism. For that matter, moral error theorists call all moral utterances false, and to do so is to admit that moral utterances have truth-functions.[4] To reject moral realism need not entail we reject cognitivism.

We have to distinguish (A) whether moral utterances can be objectively true or false in themselves, i.e., as descriptive statements, or (B) whether moral utterances can be true or false in terms of some reductive relation supervening on objective facts, or (C) whether moral utterances are non-reductively true or false in terms of their evaluative component only, or (D) whether moral utterances cannot be true or false at all. By this division, (A) expresses non-natural moral realism a la Moore and most recently expressed by Shafer-Landau.[5] (B) expresses a form of naturalized ethics, (C) expresses the Timmons-Gibbard-Blackburn cognitivist-non-descriptivist-irrealism, and (D) expresses the non-cognitive non-descriptivism, reminiscent of the emotivists. (B) can be subdivided into two further camps: B1 (realist) and B2 (irrealist) camps. Realist camps will claim that the supervenience relation is itself objective and necessary. Proponents of B1 include Paul Bloomfield, David Copp, Peter Railton, William Casebeer, and Daniel Robinson.[6] Irrealist reductionists will not ascribe any objective component to the supervenient relation. I opt for B2.

Concerning (D), one would imagine the presumption is in favour of moral utterances being truth-functional. After all, "*x* is moral" entails "'*x* is not moral' is false," which entails "'*x* is moral' is true." And although it is possible to say truly both that "*x* is moral" and "∼*x* is moral," (perhaps *x* stands for "sneezing"), one cannot truly say both that "*x* is morally required," and "∼*x* is morally required." That the logical derivations here make sense highlight the ease in which we can ascribe truth values to moral utterances. At least, so the critique against non-cognitivism goes. It strikes me, however, that the argument does not succeed. Compare the following two arguments:

(1) All and only good persons are to be admitted to heaven, and since Joe is a good person, Joe will be admitted to heaven.
(2) All and only good persons are to be admitted to heaven, and since Ned is a bad person, Ned will be admitted to heaven.

We would say that (1) is valid, while (2) is invalid. Since the difference between the two rests in the differing normative values in the second premise of each argument, those normative value judgements cannot be meaningless. The problem is that non-cognitivists will simply point out that there is also a normative term in the first premise. And therefore both arguments *are* equally valid, which can be shown by simply removing all normative terms in all propositions. Thus both arguments get reduced to:

(3) All and only persons are to be admitted to heaven, and since Joe/Ned is a person, Joe/Ned will be admitted to heaven.

Railton tries something similar.[7] Take the following argument:

(4) "If stealing is wrong, then encouraging others to steal is also wrong, and since stealing is wrong, encouraging others to steal is wrong."

Since (4) is valid, the propositions must in fact *be* propositions: that is they must be truth-functional, and thereby not meaningless. But, again, a non-cognitivist need not be bothered. For we may equally say,

(5) If stealing is a sneeze, then encouraging others to steal is also a sneeze, and since stealing is a sneeze, encouraging others to steal is a sneeze.

The validity remains without our being forced to admit the premises have meaning. Validity is defined as asserting that *if* the premises are true, the conclusion is forced upon us, and nothing in this definition entails the premises must be true or even have meaning. For that matter, we can show logical relations between geometric shapes. If a square gets us a circle and we have a square, we necessarily get a circle. So pointing out logical relations of moral terms does not show that those terms themselves have any meaning.

That said, an irrealist can accept that there are moral properties and moral facts, but reject that these properties and facts are objective. A non-cognitivist would not even accept there are properly speaking moral facts and moral properties. To non-cognitivists, I guess, a fact necessarily makes reference to some objective property. But this position seems unnecessarily narrow. Grammatical rules certainly seem to be cases of non-objective facts. But making grammatical mistakes need not entail that we conceive grammar to be a real, objective property. Similarly, the possibility of making moral mistakes need not entail moral realism.

The standard-bearer of non-cognitivism is emotivism. Emotivism is perhaps a dead theory. Non-cognitivists deny that moral utterances are capable of being either true or false. They deny that moral claims have the requisite features associated with true or false claims. Emotivists assert that moral utterances are not truth-functional and are no more than emotive ejaculations.[8] The conjunction is normally thought redundant, but Timmons and Blackburn argue for a truth-functional moral expressivist position. Modes of expression may be appropriate or inappropriate,[9] correct or incorrect.[10] That is, emotivism and expressivism are not the same. On the emotivists' view, moral claims are no more true or false than sneezes, or outbursts associated with a hockey team's scoring a goal. Moral disagreement is not over any facts, merely attitudes, and these attitudes are not themselves rationally justifiable. Sure, we may yell "Boo!" or "Yeah!" but no one has any motive to take heed of our ejaculations – even if part of our purpose in yelling "Boo!" and "Yeah!" was to persuade others to join our chant. Emotive ejaculations may persuade some people, just as associate advertising may persuade some people, but neither *ought to* as is. There may well be justification to buy a product, but that justification is not shown merely by associating it with emotively loaded images. That one ought not interfere with someone's abortion, too, may be justifiable, but not merely because a pro-choicer yells "Boo!" at abortion interventionists. To evaluate one norm over another, it appears we must speak in terms of something true about the world or about societies.

If societies really do better by adopting a rule like, "You ought not steal," what is problematic with linking the truth of, "You ought not steal," with the empirical claim concerning whether or not societies do better adopting sanctions against stealing? Whether, "You ought not steal," is true or false seems irrelevant here. It is practical or it is not, and *that* may be true or false. Consider the advice against smoking. We can say it is true that smoking is bad for you, but saying, "You ought not smoke," is not something that can be true. But nor is it something that stops merely at the emotive level of the speaker. The normative advice against smoking is connected to something about the state of the world.

For (C) versions, let us take Timmons. In the following, capitalization refers to real objects in the world existing independently of observation. Timmons notes, "[a]lthough there are no moral PROPERTIES or FACTS, property and fact talk have their proper place in moral thought and discussion.... Fact talk (and...property talk) is proper and legitimate *when made from within a morally engaged perspective*."[11] Fine, but is that perspective proper to take when it entails non-truths, namely the belief that there are moral PROPERTIES and FACTS? We know talk of God is proper from within a Christian perspective, but is it proper to hold that Christian perspective if we have antecedent reasons to believe GOD does not exist?[12]

Timmons hopes to avoid this complaint by highlighting that moral discussion is not descriptive discourse. "When I assert that apartheid is practiced today in certain countries, this assertion, if sincere, is a linguistic expression or reflection of my belief that apartheid is practiced today in certain countries; it does not *describe* me as thinking or believing this claim. So, likewise, when I sincerely say that apartheid is wrong, my assertion reflects my moral judgements (belief) that apartheid is wrong; it does not *describe* me as having such a moral stance."[13]

Fine, but does one's evaluative discourse not pretend to describe the world as being so constituted that one might say one's evaluation that apartheid is wrong is true? No, says Timmons. Moral assertions are purely evaluative. They do not describe the world; they evaluate the world. But in what sense are evaluations true or false, save that there is some connection to the states of the world? After all, to say "*x* is moral" is truth functional only to the extent that the speaker has such evaluations is to confuse Russell's notion of knowledge by description for knowledge by acquaintance. If Jones says, "The world is flat," we could say it is true that Jones believes the world is flat without committing ourselves to the claim that the world is flat.

Blackburn hopes to avoid the above charge. "Expressivism denies that when we assert values, we talk about our own states of mind, in actual or potential circumstances. It says that we *voice* our states of mind, but denies that we thereby describe them. Similarly, if we are sincere when we say that 'the time is midnight' we voice our belief, but we do not describe ourselves as having a belief."[14] But then all utterances are expressions – not descriptions – even so-called descriptive statements like, "It is midnight," or, "The cup is on the table." If the expression, "*X* is immoral," is just like the expression, "The cup is on the table," expressivism is uninteresting. I express my description of the world. I do not describe my belief about the state of the world. (What would that be like anyway? Is my belief, "The cup is on the table," a round, ochre sort of belief?)

If you prefer, consider the following normative claim: "It is good to use Phillips head screwdrivers for Phillips head screws." Why would we be tempted to call this evaluative utterance true? We would call it true only because Phillips head screwdrivers truly fit the Phillips head screws in ways that no other screwdrivers can. That is, our evaluation is based on facts of the world, coupled with one's desire to screw this particular screw. It is that description of the world coupled with a descriptive account of desires that causally connects the normative evaluation. But what is the nature of this causal connection? That the causal connection is not anywhere nearly as clear as reductionist accounts demand leads many to non-reductionism. Timmons, for example, argues that "moral sentences have a primitive kind of content: evaluative content.... In answering a question about the content of a moral sentence, the proper response is to simply repeat the sentence."[15] Moral utterances can never be fully reduced to descriptive statements of the objective world. Following Moore's Open Question Test, we can always ask, "Yes, but is it *good* to use a Phillips head screwdriver?" Descriptive reductions will, at best, give us prudence, but never moral evaluation. A reductionist natural non-realist (B2) need not worry about this charge, however. So long as we couple the facts of screws and drivers with the desire to screw this particular screw, we have all that we need. To the extent that the normative utterance supervenes on the natural facts and properties of the world, the normative utterance may be said to be "true" or "false," but to the extent that the evaluative component rests on subjective desires of agents, we cannot speak of the normative link as being objective or necessary in the sense that the moral realist demands.

Consider the task of baking an apple pie. What is an apple pie, but a conglomerate of separate bits and pieces mixed together in proper proportions and baked for a specific amount of time at a specific temperature? We can certainly say the mere ingredients alone are not enough to make the pie. The pie, in this sense, is greater than the sum of its parts, but that does not make the pie an irreducible entity completely divorced from its ingredients. Likewise with morality. Morality may be greater than the sum of the naturalistic reductionist's picture of its parts, but this does not mean the naturalistic reduction has not captured morality any more than supposing following the recipe for an apple pie cannot get at the essence of an apple pie.

The open question test fails only when one is expecting more from morality than one is entitled to get, like when a magic trick is explained without any reference to real magic. Saying $A = B$ is not necessarily false merely if the audience refuses to believe B captures what is meant by A. The audience may be mistaken, and their not being mistaken is not gainsaid by their saying they are not mistaken. We have reduced magic to non-magic, and someone may complain that something is missing. Something is missing, all right: the very thing we intended to do away with. The thing missing is – to us – unreal in the first place. We have left the acts, the things we experience, and got rid of the metaphysical fiction.[16]

Naturalist reductionist ethicists maintain that moral properties supervene on lower-level (non-moral) properties in the sense that the lower-level properties *necessitate* the higher-order so that the moral properties may be described in purely natural, non-moral terms. Supervenience and reductionism are not necessarily the same. The supervenience relation between a moral property and a non-moral property may be symmetrical or asymmetrical. If it is asymmetrical, changes at the moral level require changes at the non-moral level, but changes at the non-moral level do not necessarily entail changes at the moral level. If it is symmetrical, changes at the moral level require changes at the non-moral level, and any change at the non-moral level require changes at the moral level.[17] If the relation is symmetrical, moral properties may be said to fully reduce to non-moral properties. Such a strict identity thesis is hard to accept, however, since we can imagine many changes at the non-moral level without altering our moral stance. In fact, the idea that morality is to be impartial and universal highlights the asymmetrical relation between moral properties and non-moral properties. Moral reductionist theories demand strict symmetrical supervenience. But if the supervenience relation is asymmetrical only, it would appear that moral properties are still something beyond non-moral properties. Naturalists are against supposing anything supernatural is involved here, but moral realists need not be non-naturalists. Bloomfield, Copp, Casebeer, and Robinson, for example, take a natural moral realist stance.

Mackie, recall, rejects non-natural moral realism as being too queer, metaphysically, epistemologically, and motivationally.[18] In response, Robinson asks us to consider our reaction to a cry for help. We often know what to do. What is our knowing what to do based on? The answer is an "ensemble of postural, facial, and gestural actions that shows the desire for assistance. Thus it would be idle to ask what sort of 'sense organ' is responding to or recording the actor's plea for help. There is nothing in the transaction that is at all 'queer', though there is also nothing essentially 'sensory or particular'."[19] What Robinson says here is true, but seeing a desire for help is not the same as seeing an obligation to help. As Harman makes clear in his burning cats illustration,[20] moral statements are not entailed by description statements of fact. Mackie is certainly not talking about an inability to pick up myriad complex, sensual cues – he is talking about an *added* feature: morality – someone's saying, "You *ought to* help!"

The prospect of a natural moral realist stance is plausible so long as the supervenience demarcation is divided between symmetrical and asymmetrical lines. In comparison to the development of the mind-body debate, the choice offered us is between a strict identity thesis and a dualist thesis. Such a strict demarcation excludes a functionalist position. A functionalist symmetrical supervenience will tolerate certain deviations at the lower non-moral level without affecting the higher level moral evaluations, while remaining fully reductionistic.

Timmons denies there are any objective moral properties or facts one can appeal to in making one's moral judgements. I agree with this. Indeed, I can also agree with Timmons that "one may be epistemically responsible in holding certain beliefs without needing justification,"[21] and he wants to call such beliefs "contextually basic beliefs."[22] I would call a moral realist someone who holds there are moral properties or facts, whereas a natural reductionist irrealist claims moral discourse is shorthand for a connection to natural properties and facts and has normative force for those who stand in certain relation to those natural properties and facts via certain desires, preferences, and interests. Robinson describes moral realism as holding that "[t]here is a right answer concerning the moral thing to do, and it is independent of psychological, biological, and social contingencies."[23] I hold that there is a right answer concerning the moral thing to do, but it is dependent on psychological, biological, and social contingencies. In Timmons's sense, moral constructivists are moral realists to the extent that they admit moral properties and facts, but these are human constructions. I argue that their supervenience on objective properties and facts does not make the supervening properties objective. To be an irrealist is merely to deny the existence of objective moral properties and facts.

Timmons identifies my B2 position with constructivism. Constructivism "treats moral judgments as complex questions about how people would view things and, barring some nontrivial problems about idealized counterfactuals, answers to these questions would be capable of truth or falsity."[24] Accordingly, "*x* is immoral" means something like, "People in general would disapprove of allowing *x* as part of our basic scheme of social cooperation were we to consider the question with full information from a standpoint that considers the well-being of all affected without partiality."[25] Here there is an interesting mixture of realism and counter-realism. Given the antecedent conditions, we are appealing to real descriptive features of the world. The problem, though, is that the appeal to these real features of the world are made in the hand-waving gestures of a pseudo-reality. What we are given is an ideal under which moral assessment could be made. Nothing in accepting this commits us to imagining such an ideal is thereby an objective reality. Constructivists – so defined – have simply pushed the discussion back one notch. We can barely say that it is true that idealized rational agents under suitable circumstances would agree to $\sim x$. To do so would require our nailing down very specifically those idealized conditions, which, as Railton admits, is difficult. But even if we could say that it is true that $\sim x$ is what idealized agents in idealized circumstances would agree to, we cannot say therefore it is true that $\sim x$ is moral (and x immoral), unless we also define truth in terms of idealized agreement. We might simply stipulate that that is how we define morality, and I am not presently opposed to that. Still, *if* we define morality in terms of agreement, it is difficult to speak *also* of objective moral properties and facts. We could say, of course, that an agreement is made, but can we say it is a fact that a special *ideal* agreement *ought* to be made?

My complaint does not hinge on accepting a correspondence theory of truth, by the way. Certainly coherence and contextualist models more easily link truth with agreement. It may even be excusable to speak in terms of ideal agreement if we delimit the kinds of agreements we treat as truth-making. Following Rorty, for example, we may restrict truth-making agreement to agreement among certain experts. But this is irrelevant to my point here. My concern is with the concept of "ideal." We may well cite agreement among experts as a fact, but when we are talking about what an agent suspects an idealized agent would agree to in an idealized context, that fact disappears.

Of course, we may reasonably assume that ideally situated agents would have access to the truth. Their agreement on a proposition, thereby, would entail the truth of that proposition. But this will not help matters. All that we would be entitled to say are comments like the following: "God exists *if* ideal agents who know that God exists would agree that God exists." It is a highly uninteresting thing to say, of course. We might as well say "If God exists, God exists." It is no skin off the atheist's back to admit that if God exists, then God exists. Similarly, it is no advancement in knowledge to say ideally situated agents would agree that x is moral when we assume ideally situated agents have access to truth. If people knew that x is objectively true, we would expect they would proclaim that x is objectively true. But the reverse does not hold. People's proclaiming x is objectively true hardly shows, all by itself, that x is objectively true. Reality is not democratic, although what we take to be reality may be democratic. Someone can certainly say "It is objectively true that around here abortion is deemed immoral," without our assuming the speaker believes that abortion is objectively immoral. Likewise, when someone says, "Flatearthers think the earth is flat," we may be entitled to call that an objectively true utterance without presupposing that it is objectively true that the earth is flat, let alone that the speaker believes the earth is flat. When we are wondering whether the statement, "The earth is flat," is objectively true or not, we are not helped by pointing out that it is true that flatearthers take the utterance that the earth is flat to be objectively true. Nor do we call a statement false merely because everyone takes it to be false. They must take it to be false for reasons independent of what other people take it to be. Similarly, finding universally held, or paradigmatic, moral utterances, like "It is wrong to torture children," does not refute the irrealist position. This holds even if some of us have no good reason for taking paradigmatically false claims to be false (or paradigmatically true claims to be true), except for the fact that no one takes it to be true (or no one takes it to be false). For example, someone must have suggested that the earth is not flat initially, and he could not have maintained its truth on the grounds that others also take it to be not flat. He needed to give an argument for the earth's not being flat despite what people around these here parts took the shape of the earth to be. True, we would not have called his utterance true *until* general consensus took hold, but this confuses epistemology for metaphysics.

This should be obvious to everyone but pragmatists, but for some reason moral truth is not generally thought of in the same way. After all, the shape of the earth is not a matter of human behaviour. Moral discourse is. When people give evidence that moral utterances can be true, they cite agreement. But the mere fact that people around here take murder to be wrong does not support the claim that "'murder is wrong' is a moral fact." People may well believe it to be a moral fact, an objectively true statement, but not because people would take murder to be wrong. Those who believe "murder is wrong" to be a moral fact think it true *independently* of what people around here happen to think about murder. They may be wrong in supposing that the statement "murder is wrong" is an objective fact to be assessed independently of what people's attitudes about murder are. On this much I agree with Timmons. That kind of metaphysical certainty is in doubt.

The main argument for moral realism is by analogy to scientific knowledge.[26] David Copp provides the classic morality-science analogy.[27] Plebeian Jones claims he knows that lightning is electromagnetic energy and few challenge him on this, even though Jones personally has not made any study on that matter. Scientists have said so, no scientist (so far) is saying otherwise, and no evidence Jones has counters it, so we count Jones as justified in his belief. Since knowledge constitutes justified true belief, or reliably justified true belief, we feel as content to say Jones knows lightning is electromagnetic energy as we are to say we know a period ends this sentence. Realists can argue that moral knowledge is the same. Murder is wrong, Smith says. Others do not doubt it, though no research was done on the matter by Smith, presumably. Rather, she simply echoes the sentiments she has heard from others in society, and it fits her evidential experience. Smith has no countervailing evidence. We should therefore count Smith's belief as knowledge just as we do in the case of scientific knowledge.

Nothing in this account tells us that murder is objectively wrong, only that it is true we tend to hold murder wrong. Yes, says Copp, but the same may be said for the science case. Nothing in the reliable procedure of knowing can deductively connect to whether what we deem we know is true.[28] We may be wrong. Some further information may turn up to indicate that lightning is not best explained as electromagnetic energy. We cannot now imagine what that would be, but scientists have no qualms about being fallible about such things. So nothing prevents us from doing the same with moral knowledge.

The real difficulty with the morality-science analogy lies elsewhere. I have hinted at the problem above, and now I can make it more explicit. Even if we loosen our criterion of knowledge so that we can be said to know something merely on the grounds that others "know" it, not all knowledge is gained this way. Following Copp, let us call it the *standard-way* when regular people are entitled to hold beliefs as knowledge on the basis that a consensus of experts in a recognized and related field hold that claim. But those experts cannot be entitled to call the claim knowledge in the standard-way: they must hold it in some different way, which Copp calls the *critical-way.*[29] Scientists base their beliefs on something other than plebeian assent. In the case of scientific propositions, the critical assessment will always be in terms of empirical tests. So we discover that the standard-way is justifiable so long as it is predicated on a reliable critical-way of knowing.

One might object that Jones may be justified even if the scientists have been lax. This would be so only if there is a general consensus among the scientists, that is they are all equally lax. If the experts within a field disagree, a belief Jones holds that happens to cohere with merely one of the critical disputants in the particular field of expertise cannot make Jones's belief justified. Without the general consensus of experts within the field of expertise, laymen must withhold judgment. On both these differences, the moral case fails. No standard-way appeal to the norm can count as a parallel case to scientific knowledge without the backing of a critical-way of knowing. But we cannot count as a critical-way of knowing any knowledge claim made by a particular sub-set of the critical members, since disagreement among critical members is conceivable (in fact is often the case). In particular, disagreement among moral philosophers is the norm.[30] Who are the plebeians to determine which of these differing accounts is the one that will win out? As a result, no standard way of knowing what is moral and what is not can get off the ground. True, there is consensus on paradigm cases, like, "Do not torture babies for fun," but the critical consensus we need concerns *why* we say this is wrong. Of course, the *standard-way* of knowing need not have this kind of depth, but to make the scientific parallel stick, we need a way of making sense of moral debate at the *critical-level,* and that requires some form of consensus-inducing justification. To say we do not need consensus on the why question, that it just *is* wrong, will not do when what we are really concerned with is extrapolating the wrongness of torturing babies for fun. Once we isolate what makes torturing babies for fun wrong, we can decide the moral status of abortion, of euthanasia, of homosexuality, and the like. We want to understand the structure of moral discourse itself.

The strongest defence of naturalist reductive theories is the game theoretic evidence that moral dispositions have evolutionary fit. We can say it is true that moral dispositions have evolutionary fit without saying moral precepts are objectively true. That a thicker thorax has evolutionary fit in cold climate drosophila does not mean it is an objective truth that drosophila ought to have thicker thoraxes. It is the introduction of the "ought" claim that removes any sensible ascription of objective truth value. Moral utterances are given as normative advice. Not all normative advice is moral, of course. One might even imagine someone saying, "x is morally required, but do $\sim x$" – presumably an immoral person, or a moral person who treats "morally required" as merely what a particular error-prone cultural norm dictates. Let us imagine moral normative advice telling us to offer fair divisions when bargaining. This is not a kind of statement that can be true or false in and of itself. It may be derivatively true or false only as an extrapolation of the results of evolutionary game theory where fair bargainers have more fit than unfair bargainers in bargaining games. The normative advice can be said to be supported by a claim that is objectively true, but it does not follow that the normative advice itself is objectively true or false.

Even if (what I take to be obvious) the concept of truth or falsity cannot apply to moral utterances in and of themselves, it hardly follows that we must accept non-cognitivism in the emotivist sense. There is much room between denying that moral utterances are properly true or false in themselves (A), and claiming that moral utterances are only emotive ejaculations (with persuasive components) (D). When we call an act moral (or immoral), we want to be able to point to something about the state of the world in order that, if the same feature arises again concerning another act, consistency demands we call that second act moral (or immoral) as well.[31] As Bloomfield notes, "[g]iven the kinds of creatures that we are, at the very least, there are some environments in which we flourish and others in which we flounder."[32] Even Timmons argues that "the whole point and purpose of moral practice is action guidance in ways that promote survival-enhancing coordinative behaviour."[33] Bloomfield is a realist and Timmons is an irrealist, yet both agree on the evolutionary benefits of moral norms. For both, moral claims have some connection to states of the world.

One of Bloomfield's arguments for moral realism, for example, follows from recognizing that we have moral failings. We understand that some of these moral failings are not even necessarily known by us. We can understand what it would be like not to have certain moral failings. There is a sense, in other words, in which we recognize that we can be *better* morally. Bloomfield draws the conclusion from this that our moral characters are independent of how *we* are. That is, the very concept of moral development provides some support for moral realism.[34] That morality is independent of how *I* am, however, does not show that morality is independent of how *we* are. Compare: "I understand that I do not know all the rules of golf. I understand (sort of) what it would be like to know these rules. That is, I can understand in what sense I could know the rules better than I do. This shows Golf rules are independent of how we are. This lends some support to recognizing that golf rules have an objective realism." Of course, nothing of the sort follows. Golf rules are simply an invention that convention has come to sanction.

One may complain that golf rules, as such, are real. Conceivably those who play golf – those inside the rules of golf – will conceive rules of golf as real. After all, if the rules are not real, then so too is the game itself not real. Although all will admit the game is a convention, that hardly means they will admit the game is unreal. Specifically, if the fundamental rules are different (perhaps mulligans and gimmes are allowed), the game itself will be different – not merely in degree but in kind. We would no longer call the game "golf," or we would qualify it by some term like "new" or "North American." A similar response is available to those inside the moral game. One's admitting that moral rules are social conventions does not mean one admits moral rules are unreal. And if one changes the fundamental moral rules (perhaps we agree that theft is permissible), the result will not be a moral system. My reply is to admit that there is a sense in which speaking of the reality of rules of golf and the reality of rules of morality make sense, but that this sense cannot be consumed by the thicker sense that moral realists require. The reality comes after the invention, not before the invention. The reality of golf rules and moral rules are not discovered like we discover fossils in the rocks.

Bloomfield disagrees. For him, morality is real in the sense that it is discovered. Bloomfield's main argument for moral realism rests on an analogy between health and morality. Health and diet varies from animal to animal, and culture to culture, and even person to person, but none of this need undermine the belief that matters of health and diet are objective matters. What a diabetic should eat is not necessarily what a non-diabetic should eat, and certainly what is good for a plant is not necessarily good for a person. Healthy functioning is relative to, among other things, species, age, sex, weight, activity levels, and genetic make-up. Mere relativity, or mere contingency, need not undermine the belief that objective facts underlie our ascription of proper diet and health across these variables.[35] All that matters for realists is that the standards of health are not personal, or conventional or constructed by us. The rules are based on standards that we are not free to invent; we discover them. And they are grounded in a reality that is sufficiently independent of the judgements that we happen to make. As Bloomfield observes, "no one can be a nonrealist about health.... To think otherwise is to be committed to denying that being healthy differs from being sick, that life differs from death."[36] The problem, though, in connecting health to morality is that health concerns oneself. Showing we have a reason to abide by our self-interest in our health is not the same as showing we have a reason to abide by the interests of others. I can admit that my victim has preferences against being my victim, but such realism is not sufficient to show me the realism of a moral rule forbidding me from harming my victim.

Health is understood functionally. We might tolerate saying that a heart has the function of pumping blood (so long as this does not commit us to inferring intentionality). From the basis of that ascribed function, we can speak of a heart's being healthy if it satisfies its purpose, and unhealthy or defective if it fails to satisfy that purpose, or does it poorly compared to other hearts. If moral realism is to be analogous to the realism of health, then something's being moral must be predicated in terms of the function of the person to be moral. This does not make sense, though. Your function as a biological being may be *x, y, z*, but does being moral fit in that list? What is functioning well when an agent is being moral in the way that the heart is functioning well when pumping blood? Despite Aristotle's thinking so, it is difficult to understand the function of being human in any way comparable to the function of a heart or a hammer without resurrecting an intentional designer. And if we can describe humans as having the function to be moral, presumably we can also describe them as having the function to be rational. As I shall argue in Chapter 4, however, these two "functions" collide.

The problem with trying to link morality to health is that it commits the fallacy of composition. We are asked to move from the purpose of a morphological feature, like a heart, to the purpose of a whole organism like a human being. That a morpohological feature of an organism has a function does not mean the organism as a whole has a function. That a tree serves a function within the ecosystem does not give a reason to believe the ecosystem as a whole has a function. One may try to save the case by speaking in less universal terms. Given that the function of the brakes are to stop the car, if they fail that function, we are entitled to say the brakes ought to be fixed. Similarly, Bloomfield argues, we can say, "A borrower ought to return what she borrowed," which he translates as "It is the function of a borrower to return what is borrowed."[37] Fine, but this amounts to little more than saying, "It is the function of a moral person to be moral," which is not the same as saying, "It is the function of a person to be moral." Surely it is the latter that realists needs to say, since the former is simply uninteresting. Realists want to say, "The purpose of an agent is not to steal," but all they are entitled to say is, "The purpose of someone committed to not stealing is not to steal." Hardly interesting. For that matter, would we even say that "the purpose of a borrower is to return what is borrowed"? Is that the purpose of a borrower? I would imagine we would say that the purpose of a borrower is to get the use of an object on loan. We cannot link health realism to moral realism.

I do agree that nature has something to tell us about morality, and I shall attempt to clarify this in evolutionary terms. Simply, I resist that this has anything to do with moral realism. When I refer to evolution, I do not mean to imply anything connected to a telos – as Casebeer and Robinson suppose.[38] In fact, I mean the opposite. There is a direction to evolution, certainly, but there is no built-in normative notion to that direction. A species chugs along until it dies out and that's pretty well it.

1.3. The Scope of Morality

We need a way of distinguishing moral normative imperatives from non-moral normative imperatives. The distinction lies in whether we are dealing with interpersonal preference conflicts or not. I hold that morality's role is to offer resolution to interpersonal conflicts over preferences. This is not what people generally understand morality to be about, but I shall argue it follows from viewing morality as an evolutionary strategy. If there were no conflict, morality would not evolve. If the conflicts of interests occurred only *within* the person, again, morality would have nothing to say about that, assuming no repercussions to others is expected from either satisfying or not satisfying either of the internal interests. Let us call the strategic interaction view of morality the *narrow* view. How one acts in relations to others may be a moral matter, whereas it is never a moral matter, on the narrow view, when your actions concern yourself alone. In this sense, morality's role is not concerned to satisfy preferences, since some preferences people have concern themselves alone. This distinction will be important in Chapter 3. An alternative picture is offered by a *broad* view of morality. Proponents of a broad morality (let us call them "broaders") believe that intrapersonal goals ought also to play a part in morality in a way that is not merely derivative of interpersonal goals. Broaders believe morality extends also into the intrapersonal domain.[39] According to Bloomfield, the subject matter of moral discourse is constituted by both "intrapersonal and interpersonal relations."[40] And Charles Taylor holds that any deliberation that requires what he calls "strong evaluation," will be a moral matter.[41] Strong evaluation consists in our distinguishing right from wrong, good from evil, noble from base, and other sorts of contrasts. Accordingly, any desire that involves strong evaluation is necessarily a moral decision. Since we distinguish right and wrong actions even in our intrapersonal spheres of life, morality must necessarily be "broadly" construed.[42]

I have no qualms that we do deem some intrapersonal desires worthy of who we are, and others unworthy, and one way to articulate this contrastive feeling is to call one "good," the other "bad." What needs to be argued, however, is why the mere use of these otherwise "moral" words is enough to draw the matter into the moral purview without simply equivocating on terms. Typically we believe that not all recommendations are moral ones. For example, "Use a soft toothbrush if you do not wish to harm your gums," or, "Don't watch too much TV," or, "Do something more with your life than count grass blades," may all be good recommendations, but none are moral recommendations on the narrow view.

If one continually beats up on people, then an individual goal to control one's aggressive urges would be a moral matter on both the narrow and broad view. Broaders, however, demand something stronger: individual goals that have no influence on one's interpersonal goals also matter morally.[43] By contrast, narrowers point out that people have disparate views of what the good is. Since conceptions of the good will tend to differ, conflicts over what one ought to do will likely arise in interpersonal relations. Morality, then, is strictly an attempt to resolve these interpersonal conflicts. Where no interpersonal relation is, one's conception of the good is all that matters, and so there is no need for an arbiter ("morality" for narrowers). Once one accepts the basic premises of subjectively defined notions of the good, "narrow" ethics would seem to follow. If morality also seeps into the intrapersonal domain, as the broad view advocates, then morality limits one's personal thoughts, ambitions, and goals independently of their effects on others. Such restrictions need to be motivating to the restricted agents. That is not easily forthcoming. Its effects on others obviously cannot be invoked as a reason, for to do so would be to fall into the camp of the narrowers. But any appeal concerning the effects on the agent herself may succeed on a prudential level, but will fail to qualify as an obligation. That smoking may cause lung cancer is a good reason not to smoke, but cannot count as *obligating* anyone not to smoke. Perhaps one may "obligate oneself" not to smoke, or to do sit-ups, but just as easily, one may obligate oneself to smoke, or to refrain from doing sit-ups. Intrapersonal advice can only count as advice. And this complaint ignores the problems inherent in determining whether any intrapersonal advice counts as *good* advice, given the enormous disparity on what counts as a good life. As Sartre remarked in *The Age of Reason*, "Everybody always had Boris's good in view. But it varied with each individual." On this point, I agree with Mill: "All errors which he is likely to commit against advice and warning are far outweighed by the evil of allowing others to constrain him to what they deem his good."[44]

No one need deny that many of our most important decisions occur outside the interpersonal arena. What career I should pursue or how I should best spend my free time or whether I should believe in God are matters that concern me deeply, and yet are matters about which the narrow view of ethics could care less (unless externalities exist). Narrowers do not care about these intrapersonal matters not because they are insensitive brutes, but because they recognize that people differ as to what counts as intrapersonally important. What matters morally for the narrow view concerns only that area of "overlapping consensus," to use Rawls's terms.[45] An overlapping consensus is whatever reasonable people can agree on. To claim that matters beyond this consensus ought to be politically enforced is to impose the values of a sub-group on everyone. On the narrow view, no society has a right to demand of its populace more than what satisfies this overlapping consensus. It is not a necessary condition that an overarching consensus is minimal. It may be that everyone in a closed society believes that it is everyone's duty to profess Catholicism at every level of society: in schools, in courtrooms, and in the definition of marriage, for example. The overlapping consensus then would permit the church to become the state. Appeals to psychological realism, however, plus the desire to peacefully associate with more than merely one's own cultural group, plus the fact that few, if any, societies are sufficiently "closed," all preclude such thick conceptions of the overlapping consensus.

Basically, broaders conflate morality with well-being. Broaders hold that the "subject matter of morality concerns the difference between acting and living well and badly."[46] Narrowers separate the two. For narrowers, morality concerns acting well or badly, only. On the narrow view, morality will carve a space for the grass-counter to pursue her desires, even if we find such desires not conducive to her well-being.[47] I do not mean to imply that narrow ethics is the starting position. Rather, narrow ethics is an implication from understanding morality as offering advice for resolving interpersonal conflicts over preference satisfaction – a position to which proponents of naturalized ethics are led.

Another way of understanding the narrow-broad distinction is sometimes offered, which highlights more the distinction between positive and negative duties. The typical complaint is that morality clearly demands we help victims, prevent starvation, assist our neighbours and the like, and narrow ethics fails to accommodate these intuitions. Any moral theory that fails to include benevolence is surely "morally monstrous," if not "evil"![48] Annette Baier argues that relying on mere principles of consent cannot account for the various virtues such as "kindness," "generosity," "gentleness," and "humility." To artificially truncate morality to exclude these virtues is to offer not only a narrow, but also a distorted view of moral practices. But a strict demarcation between narrow and negative, broad and positive fails. Broaders may be more unwilling than narrowers to suppose negative rights always trump positive rights, but narrowers need not be opposed to positive rights. Narrowers merely attempt to explain why we term some acts moral and others immoral without appealing to personal well-being. Morality is explained or justified on the basis of non-moral propositions in terms of interpersonal conflict resolution. Such reductionism, many argue, is doomed from the get-go, beset as it is by bias and ad hoc rationalization.[49] Any attempt at moral justification is an attempt to reduce the non-reducible.[50] Gass put it succinctly: "[P]rinciples . . . frequently get in the way of good sense."[51] Gass does not appeal to brute intuition of the Mooreian brand. Rather, he wishes to draw our attention to the bare moral transparency of some acts. It is wrong to burn someone alive for no reason. To ask, "Yes, but what is wrong with burning a man alive?" is to suffer "from a sort of *folie de doute morale*."[52] Any reductionist theory that attempts to explain just what it is about the act that makes it wrong will be more inscrutable than the event it attempts to explain. It will sound simply ludicrous, Gass avows. Here I refer back to the reduction of magic. That a conjurer's trick is explained without recourse to "real" magic is exactly what we are after. Likewise, to explain moral norms absent any appeal to moral presuppositions is a success, not a failure.

Rather than starting with principles and developing a moral or political structure from them, broad anti-reductionists argue for beginning with clear cases of morality and immorality and to treat these as data.[53] The narrow reductionist approach, however, does much the same. Simply, from these clear cases, an underlying principle is deduced. The merit of the reductionist approach, then, is to find an underlying principle that explains why we treat certain cases as moral and others as immoral. And knowing this will enable a normative theory to be developed.

1.4. Consent Theory

The betting analogy offered in my introduction is to highlight that morality cannot be rationally defended (an argument I will make more explicit in Chapter 4), but that it has evolutionary fit nonetheless (an argument I defend in Chapter 5). But what does this tell us in terms of moral advice? In Chapter 6, I argue that a similarity exists between the successful strategies in various games, and this will have something to do with conditional cooperation. From this, we can extrapolate the following abbreviated normative rule: *Do not do unto others without their consent.* I shall call this the *principle of consent.* At first view, the principle of consent will not appear to be a conditional strategy, but any other player, *B*, who rejects the principle of consent releases any player, *A*, from any wrongdoing toward *B*. By rejecting the principle of consent, *B* consents to any act done to her. In turn, this shows the merit of accepting the principle of consent. But, given the complexity of strategic interaction, such an argument falls short of providing rational justification for *B* to accept the principle of consent. The normative rule is neither objectively true, nor rationally fully justified, nor broad in scope.

Notes

[1] The following insight I owe to Daniel Dennett, who applies a similar analogy in relation to consciousness in his "Real Consciousness, Real Freedom, Real Magic" (Julian Jaynes Lecture, University of Prince Edward Island, October, 2003).

[2] Peter Railton, "Moral Realism: Prospects and Problems," in Walter Sinnott-Armstrong and Mark Timmons (eds.) *Moral Knowledge? New Readings in Moral Epistemology* (Oxford: Oxford University Press, 1996), 52, 59.

[3] Mark Timmons, *Morality Without Foundations* (Oxford: Oxford University Press, 1999), 144. Allan Gibbard, *Wise Choices, Apt Feelings: A Theory of Normative Judgement* (Cambridge, Mass.: Harvard University Press, 1990). Simon Blackburn, *Ruling Passions* (Oxford: Clarendon Press, 1998).

[4] Richard Joyce, *The Myth of Morality* (Cambridge: Cambridge University Press, 2001), 8–9.

[5] G.E. Moore, *Principia Ethica* (Cambridge: Cambridge University Press, [1903] 1986). Russ Shafer-Landau, *Moral Realism: A Defence* (Oxford: Oxford University Press, 2003).

[6] Paul Bloomfield, *Moral Reality* (Oxford: Oxford University Press, 2001). David Copp, *Morality, Normativity, and Society* (Oxford: Oxford University Press, 1995). William Casebeer, *Natural Ethical Facts: Evolution, Connectionism, and Moral Cognition* (Cambridge, Mass.: MIT Press, 2005). Daniel Robinson, *Praise and Blame: Moral Realism and Its Applications* (Princeton: Princeton University Press, 2002).

[7] Railton, *Moral Knowledge?*, 59–60.

[8] See Gibbard, *Wise Choices, Apt Feelings*, and Timmons, *Morality Without Foundations*.

[9] For example, Blackburn, *Ruling Passions*, 59.

[10] For example, Blackburn, *Ruling Passions*, 54.

[11] Timmons, *Morality Without Foundations*, 160, my emphasis.

[12] Or does God's non-existence only count from outside religious discourse, in which case it could never count as a criticism of religious discourse? If so, hope of communication dissolves.

[13] Timmons, *Morality Without Foundations*, 144. See also Gibbard, *Wise Choices, Apt Feelings*, 105.

[14] Blackburn, *Ruling Passions*, 50.
[15] Timmons, *Morality Without Foundations*, 148.
[16] David Brink rejects my associating belief in moral facts with belief in magic. Magic does not cohere with natural beliefs/facts, whereas moral beliefs do. (David Brink, *Moral Realism and the Foundation of Ethics* (Cambridge: Cambridge University Press, 1989), 183.) Similarly, Railton argues that while magic (like being cursed, or being blessed) does not supervene on natural properties, morality does supervene on natural properties. (Railton, 65.) Brink gets the association wrong. Magic does cohere with other beliefs in magic, just like belief in moral facts coheres with other beliefs in moral facts. The disagreement between me and Railton concerns, firstly, whether it is morality or prudence that supervenes on natural properties, and secondly, if it is prudence, whether our job of reduction is done. Concerning the second, I say, "Yes." Railton and Brink say, "No."
[17] See, for example, Bloomfield, *Moral Reality*, 44.
[18] J.L. Mackie, *Ethics: Inventing Right and Wrong* (Hammondsworth: Penguin Books, 1977), 38–42.
[19] Robinson, *Praise and Blame*, 29.
[20] Gilbert Harman, *The Nature of Morality* (New York: Oxford University Press, 1977), 4–5.
[21] Mark Timmons, "Outline of a Contextualist Epistemology," in Walter Sinnott-Armstrong and Mark Timmons (eds.) *Moral Knowledge? New Readings in Moral Epistemology* (Oxford: Oxford University Press, 1996), 305.
[22] Timmons, "Outlines," 297.
[23] Robinson, *Praise and Blame*, x.
[24] Railton, *Moral Knowledge?*, 70.
[25] This is Railton's account, which he takes to be consistent with cognitivism. Railton, 69.
[26] For Paul Churchland, for example, moral knowledge is used to help us navigate our social surroundings, while science is used to help us navigate our physical surroundings. Paul Churchland, "Toward a Cognitive Neurobiology of the Moral Virtues," *Topoi* 17 (1998).
[27] See also David Brink. Brink defends moral realism by defending coherentism. Beliefs are true if and only if they cohere with our other beliefs. In this sense, moral beliefs can count as facts, since they cohere with our other moral beliefs, in the same way as we confirm scientific theories. (Brink, *Moral Realism*, 125–30.)
[28] David Copp, "Moral Knowledge in Society-Centred Moral Theory," in Sinnott-Armstrong and Timmons (eds.) *Moral Knowledge?*, 262–4.
[29] The distinction between standard and critical ways of knowing is raised by Copp, *Moral Knowledge?*, 263.
[30] See Alan Gilbert, *Democratic Individuality* (Cambridge: Cambridge University Press, 1990), for a reply to my claim that there is no moral consensus at the critical level.
[31] See, for example, Richmond Campbell and Jennifer Woodrow, "Why Moore's Open Question is Open: The Evolution of Moral Supervenience," *The Journal of Value Inquiry* 37 (2003): 353–72.
[32] Bloomfield, *Moral Reality*, 19.
[33] Timmons, *Without Foundations*, 172.
[34] Bloomfield, *Moral Reality*, 12.
[35] Bloomfield, *Moral Reality*, 35–7.
[36] Bloomfield, *Moral Reality*, 38.
[37] Bloomfield *Moral Reality*, 135.
[38] Casebeer, *Natural Ethical Facts*, 43, 49–50, 71. Robinson, *Praise and Blame*, 16.
[39] Michael Sandel, *Liberalism and the Limits of Justice* (New York: Cambridge University Press, 1982). Larry Haworth, *Autonomy: An Essay in Philosophical Psychology and Ethics* (New Haven: Yale University Press, 1986). David Griffin, *Well Being: Its Meaning, Measurement and Moral Importance* (Oxford: Oxford University Press, 1986). Alasdair MacIntyre, *Whose Justice? Which Rationality?* (Notre Dame, Ind.: Notre Dame University Press, 1987). Michael Walzer, *Interpretation and Social Criticism* (Cambridge: Harvard University Press, 1987). Charles Taylor, *Sources of the Self: The Making of Modern Identity* (Cambridge: Harvard University Press, 1989). David Wiggins, *Needs, Values, Truth* (Oxford: Oxford University Press, 1998).
[40] Bloomfield, *Moral Reality*, 18.

[41] Charles Taylor, "What is Human Agency?" in *Human Agency and Language: Philosophical Papers I* (Cambridge: Cambridge University Press, 1985), 16.
[42] Taylor, *Sources of the Self*, 4.
[43] One could, I suppose, believe that solely intrapersonal goals might matter morally so long as one held a theological perspective. In this case, "Goodness" is not defined by interaction with mere mortals, but by one's interaction with a deity. In this sense, morality is still determined by interpersonal relations, just not any human relation, and perhaps not any real relation, either. Perhaps not all broaders are theists, but all theists are likely to be broaders.
[44] John Stuart Mill, *On Liberty* (Indianapolis, Ind.: Hackett Publishing Co., [1859] 1978), 75.
[45] John Rawls, *Political Liberalism* (New York: Columbia University Press, 1993), 10.
[46] Bloomfield, *Moral Reality*, 18.
[47] The grass-counting example stems from John Rawls, *A Theory of Justice* (Cambridge, Mass.: The Belknap Press of Harvard University Press, 1971), 432–33, but see also Derek Parfit, *Reasons and Persons* (Oxford: Oxford University Press, 1984), 499–500; Griffin, 323, nt. 29; and Malcolm Murray, "A Catalogue of Mistaken Interests: Reflections on the Desired and the Desirable," *International Journal of Philosophical Studies* 11/1 (2003): 1–23. For similar examples and discussion, see David Brink, *Moral Realism and the Foundations of Ethics* (Cambridge: Cambridge University Press, 1989), 227; Susan Wolf, "Happiness and Meaning: Two Aspects of the Good Life," in E. Frankel Paul, F. Miller, Jr., and J. Paul (eds.) *Self-Interest* (Cambridge: Cambridge University Press, 1997), 211; Thomas Scanlon, *What We Owe to Each Other* (Cambridge, Mass.: Belknap, Harvard University Press. 1998), 89–90; and Joseph Raz, *The Morality of Freedom* (Oxford: Oxford University Press, 1988), 301, 307–8, 316.
[48] This is Kai Nielsen's pronouncement in his "Capitalism, Socialism, and Justice," in T. Regan and D. Van DeVeere (eds.) *And Justice for All* (Totowa, N.J.: Rowman & Allenheld, 1982), 264–86.
[49] Alasdair MacIntyre, *After Virtue* (Notre Dame, Ind.: University of Notre Dame Press (2nd edn) 1984). Sandel, *Liberalism and the Limits of Justice*. Charles Taylor, *Human Agency and Language*. Simon Blackburn, "Moral Realism," in J. Casey (ed.) *Morality and Moral Reasoning* (London: Methuen, 1971).
[50] A.J. Ayer, *Language, Truth, and Logic* (New York: Dover Publications, 1952), 102–20. Charles L. Stevenson, "The Emotive Meaning of Ethical Terms," *Mind* 46 (1937).
[51] William Gass, "The Case of the Obliging Stranger," *The Philosophical Review* 66 (1957): 203.
[52] Gass, *The Philosophical Review*, 197.
[53] I should note that Rawls attempted a happy medium between these extremes that might interest those convinced by the Gass-MacIntyre lines of attack. John Rawls, *Theory of Justice*, 48–51.

Chapter 2

AGAINST MORAL CATEGORICITY

2.1. Categoricity

Morality is a social convention, nothing more. It is a very useful convention, but still a convention: a contingent notion, a relative notion, a hypothetical construct. Many disagree. They note that when speaking morally, we want to say, "Don't torture people," not, "Don't torture people if you don't want to be cruel." If for whatever reason one prefers the latter way of speaking, we would be obliged to add, "Then don't want to be cruel!" That is, it is hard to understand how morality can be anything but categorical. To ask on what this categoricity is based is to already misunderstand the demand that morality be categorical. To say the categorical nature of morality is based on x, is to imply that morality is conditional on x, but the categoricity of morality precludes morality being conditional on anything. It is not a fully moral (i.e. categorical) utterance to say, "If x obtains, don't be cruel." Categoricity (i.e., morality) wants to demand, "Don't be cruel whether x obtains or not." For if $\sim x$ obtains, what then? Who but terrorists want to say the prohibitions against cruelty may be waived if the right conditions obtain, and then point out the existence of those conditions. To put it bluntly, those who maintain that morality is not properly speaking categorical are not merely misguided, but immoral. All fine sentiments, but for one problem: categoricity is nonsensical.

Showing that a concept C is nonsensical is problematic, since C proponents will complain that I am not speaking of C, but C', where C' is my own bastardized version of C. Since concepts are large and many-faceted, this manoeuvre is imminently sensible. To be clear about what categoricity demands, let us review what Kant has to say about it. For Kant, a categorical reason for action must be a reason for all.[1] Kant insists that we can have reasons for recommending only those principles of action which could be adopted by all concerned, whatever their particular desires, social identities, roles or relationships.[2] In that a moral law must be categorical rather than hypothetical, Kant means that a law must command unconditionally rather than depend upon the adoption of some antecedent and optional end.[3] Happiness, or indeed any end other than the good action itself, is too indeterminate an end to give rise to such an imperative.[4] For Kant, "the categorical imperative is limited by no condition."[5] Importantly, "the categorical imperative is an a priori synthetic practical proposition."[6] By that, Kant means any categorical imperative "is absolutely, though, practically, necessary."[7] From such talk, we may say that categorical rules are universal, impartial, inescapable, unignorable, unconditional, non-contingent, necessary, absolute, good-in-themselves, and a priori.

One other feature of the categorical imperative, at least according to Kant, is that it is unique to morality.[8] If categoricity is unique to morality, discovering categorical-like features in non-moral imperatives would either be impossible, or entail that moral categoricity remains unidentified. This insight may be attributed to Foot.[9] Showing that categoricity is a common element in our lives seems to bode poorly for the position that categoricity is nonsensical, but if proponents of moral categoricity maintain that morality is uniquely categorical, we may rule out discussing features common to non-moral norms. If a particular feature remains after this whittling process, we can then examine that aspect in isolation. Alternatively, no feature remains after our whittling down. Like peeling away the secondary qualities of a carrot, we are not left with the primary quality of a carrot, the carrot-in-itself; we are left with nothing at all. This shall be my contention.

On the other hand, if neo-Kantians reject the uniqueness condition, Foot's discussion about non-moral categoricity is a red herring. To defend Foot's approach here, then, means we must show why the uniqueness condition is needed for neo-Kantians. After all, if we want to distinguish Homo sapiens from other animals, it will do no good to point to Homo sapiens traits that occur in non-human animals. Similarly, if we wish to distinguish moral imperatives from other kinds of imperatives, the traits that are common to both cannot be relevant. So something about moral categoricity must be unique to morality. In saying this, mind you, I am not supposing neo-Kantians – or Kant himself, for that matter – maintain that moral norms can be identified by their formal features alone, such as categoricity. The content of moral norms will be different from the content of non-moral norms, even if both – or neither – are categorical. All I am concerned with in this chapter is to explain that, whatever makes moral norms *moral*, categoricity has nothing to do with it. (Later, I shall argue that what is unique to morality is not the internal structure of moral utterances, but that morality can be understood as statistically successful (though solely hypothetical) rules concerning strategic interpersonal relations.)

2.2. Rules of Golf

Foot appeals to etiquette and club rules in her demonstration that categoricity is not unique to morality. For reasons that will come out later, let us switch the example to golf. The rules of golf are many and varied and complex and most golfers do not know even a small fraction of them, and even those rules that are known are not generally honoured by the average golfer.[10] For example, it is common that golfers adjust their lies, take gimmes, take mulligans, and ignore lost ball rules, not to mention their oblivion concerning golf etiquette. Perhaps these rule-violators do so for reasons that are internal to the game. For example, strict adherence to the rules slows down play. More likely, however, is that the rules seem excessively penal for an already overly penal game. Golf is difficult enough without following all the rules. After all, one is supposed to be having fun. Take, for example, the rule about grounding one's club in a hazard. If someone accidentally drops his club in a sand trap, perhaps he stumbles and the club falls out of his hand, this is to count as a one stroke penalty, or a loss of the hole in match play. Few non-professional golfers would even know that a rule has been violated, let alone honour the rule. Not only was it accidental, it did not provide any unfair information to the average golfer. Many would have no qualms about admitting the rule is excessive (particularly in this circumstance), and thereby ignore it. But saying this is not to say that the rule about grounding one's club in a hazard does not apply. The rule is explicit in stating it does apply. Rather, we say, we are not particularly moved to honour the rule, *even though it still applies.*

Admitting that the grounding of the club in a hazard rule does not move us is not to say that the rule applies only so long as one is moved by it. The rule applies whether one is moved by it or not. That is the rule, after all. Once we admit this, we can admit that the grounding of the club in a hazard rule applies independently of one's interests, or one's desires, or the consequences, or even, remarkably, one's intentions. That you did not intend to ground the club, that it was an accident, is irrelevant to the application of the rule and the penalty. So long as we understand categorical imperatives to be imperatives that apply independently of one's interests, or one's desires, or the consequences, we can admit that the rules of golf can be categorical. Admitting this is to admit that Kant was wrong when he claimed that only morality is a categorical imperative.

2.3. Escapability

An initial objection is possible and familiar. It goes along the following lines: True, we may say that it is "wrong" to ground one's club in a hazard, or it is wrong for a runner who has been tagged between bases to refuse to go to the bench, or it is wrong to move rooks diagonally and bishops horizontally, but in none of these cases do we say it is categorically wrong. Rather, we say is it wrong only within the confines of the particular game. To follow Searle's language, the categoricity of the rules are constitutive of the particular game or practice.[11] Within that practice, the rules are categorical, but there is no categorical rule stipulating one must engage in that practice. When I dust my chess board, I may move my bishop horizontally, not diagonally. When I am on my way home from a trip to the store and a stray ball rolls up against my heel, I am still allowed to go home. When I am not playing golf, I can ground my club in hazards all I want. Whereas morality is not supposed to have such an easy escape.[12] While there are plenty of occasions of not playing chess, or not playing baseball, or not playing golf, there is not supposed to be any occasion where one is not playing the moral game. Morality applies always. Not only are moral rules categorical within the practice of morality, there is also supposed to be a categorical rule stipulating you engage in that moral practice. We will excuse those who do not play golf; whereas we cannot excuse those who do not play the moral game. No one is free from the moral rules, whereas few are bound by the rules of golf. With the rules of golf, one may elect to be bound by them (an implicit agreement, if you will, by teeing up the ball). With morality, there is no comparable option. In other words, moral rules are not like rules of golf. It adds nothing new to say of moral rules that they operate "only so long as you are playing the moral game." There is no case of not playing the moral game.[13] But there are cases of not playing baseball, of not playing chess, of not playing golf. That is, there is some sense in which morality is *inescapable*, whereas the rules of baseball, golf, and chess are not. Thus the definition of categoricity must not be merely its non-predication on interests and consequences, but its inescapability. To claim that morality is a categorical imperative, whereas rules of baseball, chess, golf, and other conventions are hypothetical, is to claim that you are bound by the rules of baseball only so long as you play the game, and nothing dictates you play the game. Whereas in the case of morality, there is no sense in which it is merely up to you whether you want to be moral today. Morality binds us all inevitably, inescapably, categorically. At least that is what categoricists would have us believe.

But is this type of objection adequate? One might wonder why one cannot opt out of morality. Don't people do it all the time? That they ought not can only make sense from *within* the moral game. It cannot make sense from without, any more than following rules of etiquette can make sense from without. But this is not the reply Foot would give. She would complain that the objection misses her point. Part of Foot's argument was that the inescapability of morality cannot be the distinguishing feature of morality. For example, a club rule may stipulate that we cannot bring ladies into the smoking room. If one quits the club, one can bring ladies into rooms where people are smoking. Therefore the rule is said to be hypothetical only, and thereby *escapable*. But is it really escapable? One *still* cannot bring ladies into the smoking room of the club even if one quits, so the rule is inescapable for all that. When an ex-member brings a lady into a room where people smoke, he is not violating the club rule so long as the smoking room into which he has brought the lady does not belong to the club. He could have done that even while still a member of the club. So his doing that cannot count as an instance where he has escaped the Draconian clutches of the club rules.

Similarly, we could say, that everyone *is* bound by the rules of golf, whether they know it or not. Despite Jesus's never having played golf, he was still bound by the rule that he ought not ground his club in hazards. So long as Jesus was never in a hazard as defined by the rules of golf, he has never violated the rule concerning grounding one's club in a hazard. Similarly the rule, "Pay one's debts," applies even if one incurs no debts, and the commandment, "Thou shalt not covet thy neighbour's ass," applies to even those whose neighbours have no asses.

2.4. Weak and Strong Categoricity

But the mere fact that rules of golf may be understood as categorical does not necessarily show that Kant is wrong: it may instead show that Kant's notion of categoricity has some other feature, at least if we wish to hold that moral rules are uniquely different than rules of golf. Foot notes that rules of etiquette apply whether or not a boor has a reason to follow them. Rules of golf apply whether or not a non-golfer has a reason to follow them. So in this sense the rules remain categorical. It is not that they apply only if you have a reason to follow them: they apply in any event. Of course this sort of categoricity – call it a *weak* categorical, after Joyce[14] – is not quite what Kant had in mind when he speaks about moral imperatives.

Let us distinguish weak categorical oughts from strong categorical oughts in terms of an all-things-considered sense. Although the rule against grounding one's club in a hazard is a weak categorical ought, it is not an all-things-considered ought. All things considered, a person may have no reason to abide by the rule. That is, persons may have good all-things-considered reasons for ignoring weak categorical oughts. Strong categorical oughts, on the other hand, must be an all-things-considered ought.[15] Moral speakers are not apt to mean that although the sanctions against killing still apply, a would-be murderer has no all-things-considered reason to bother with them. Moral speakers can readily admit that non-golfers and golf hackers can flagrantly disobey rules of golf to their hearts' content, but they will not be allowed to flagrantly disobey moral rules. We can say without contradiction the following: "Grounding the club in a hazard is against the rules of golf and so according to the rules of golf, one ought not ground one's club in a hazard, but this applies only for those who care about the rules of golf." Can we say the following? "Murder is against the rules of morality, and so according to the rules of morality, one ought not murder, but this applies only for those who care about morality." We cannot add that one ought to care about the rules of golf, but we do feel entitled (rightly or wrongly) to add that one ought to care about morality. Morality entails a strong categoricity, not merely a weak, or Footian, categoricity.

Strong categoricity is harder to justify, however, and this is where the problem begins. The strong categorical ought in the utterance, "One ought to care about morality," may be as indefensible as maintaining that even non-golfers and pseudo-golfers ought to care about the rules of golf. This is unpalatable to categoricists, but palatability is not a criterion for truth. Consider an abhorrence to not having one's teeth brushed. Perhaps not everyone will have such an abhorrence, but for those of us who do, we cannot imagine failing to do so. Certainly we may miss a brushing, perhaps two or more under imaginable circumstances, but to never brush one's teeth is unfathomable. In other words, the normative force to brush one's teeth is strong, categorical one might say. It is not that one ought to brush one's teeth if one wants healthy teeth and wants to avoid bad breath, but that one ought to want healthy teeth and avoid bad breath. It is one thing to have bad breath and bad teeth, another to actually desire such things. Our indignation toward someone who would desire such things would be strong. If the normative force for brushing one's teeth is a cultural inculcation, however, the strong categorical feel may be maintained without admitting a strong categorical reality.[16]

2.5. Arbitrariness

Categoricists may argue that morality has a different purpose than the conventions of baseball, chess, golf, social clubs, and etiquette,[17] and that Foot missed this. For example, why is it the case that there are three strikes, but four balls? Why ought we incur a penalty merely for dropping one's club in a hazard? If en passant is permissible, why can we not capture a rook traversing the knight's home square on a queen-side castle? Why must the fork be on the left and the knife on the right of a place setting? How we answer these questions are different than how we answer similar questions concerning moral rules. In the case of baseball, golf, chess, and etiquette, we end up answering the questions by repeating that those are the rules.[18] That is, we can admit that the rules of baseball, golf, chess, clubs, and etiquette are fundamentally arbitrary without affecting our participation in those games. If the explanation for moral rules is likewise arbitrary, this would very much impede our adherence to morality.[19]

If our decision to punish transgressors of a particular act is arbitrary, we have prudential reasons to abide by the arbitrary rule (to avoid punishment), but the sense that the rule is *moral* will have vanished. Being punished for something that is arbitrary is the *sin qua non* of injustice. We would not tolerate it in morality, yet such arbitrariness of rules in golf, chess, baseball, and etiquette is both tolerable and routine. The reason for this difference, presumably, lies in the fact that the penalties of rule violations in golf, in chess, or in baseball have no effect on our lives except within the confines of those activities. As Foot has shown, we cannot escape the rules of golf or the club rules, but that is not the deciding factor. We can escape the *impact* of the rules should we so choose. That is what we cannot do in the case of morality. This shows a fundamental difference between the rules of golf, for example, and moral rules. That one is bound by a rule in Foot's weak categorical sense does not necessarily mean it has much impact on one's life. That Jesus was bound (weakly) by rules of golf would hardly worry him. The same cannot be said for morality.

If the rules of golf or baseball are deemed fundamentally arbitrary, we are not typically bothered, since the reason for playing is found in the game as developed. Its historical making (however interesting) is irrelevant. The game is enjoyed as it is or it is not. If it is not enjoyed, we would not likely play it. We do not care how the sport originated – we play because it is enjoyable to us now as it is. Nor would we seek a divorce merely on the grounds that the initial meeting of our spouse was an arbitrary fluke. To discover that a moral sanction is based on some arbitrary fluke, however, is a reason to remove the sanction.[20] In other words, it should strike us as problematic if we cannot justify the imposition of the moral rule beyond arbitrary whim, or an appeal to the tradition of doing-things-that-way-around-these-here-parts. Those defences, on the other hand, seem both permissible and routine for golf, baseball, chess, etiquette, and club rules. And *that* is the difference – so categoricists may argue.

But it is unclear that we would be bothered by discovering that the origin of morality is no less arbitrary than the origins of golf. By saying this, however, I mean to distinguish two types of arbitrariness. On the one hand, the charge of arbitrariness is used in tandem with the charge of inconsistency. If rules are inconsistently applied, we may complain about arbitrariness of the application of the rules, and by this charge we normally mean unfairness. After all, allowing only some students to hand in late assignments without penalty may not be technically a case of arbitrariness: perhaps the teacher allows students she likes free extensions, but not those she does not like. The charge of arbitrariness here is really a charge of unfairness, a violation of the rules concerning extensions. Morality surely ought not be arbitrary in this sense, but nor are rules of golf, baseball, etc., arbitrary in this sense. The application of the rules may be inconsistently or unfairly or arbitrarily applied, but that can happen in the case of policing moral rules as well.

We can concede that murder is wrong and murder is understood as killing innocent persons for purposes other than self-defence. Conceding the above, we can say killing an innocent person for profit is wrong. And this follows as easily as we may say a person is "refreshingly spontaneous," a wine "full-bodied," a legal case "sloppily put together."[21] In Putnam's sense, the role of discourse ethics is precisely to come to an agreement on how to apply moral terms so that consistent application may ensue.[22] But so what? A conclusion validly derived from a premise does not make the conclusion true. If the origins are arbitrary, the end derivations have not escaped the arbitrariness. That morality is consistently applied does not show that moral rules are non-arbitrary.

The arbitrariness of social conventions, then, must be distinguished from the mere accusation of inconsistency. Rather, it is to highlight the peculiar genesis of social conventions. The queen may move like a rook or like a bishop, but never like a knight. It might have been the case, however, that the queen could also have been allowed to move like a knight, or like a knight or a rook, but never a bishop. But we do not think that murder might have been allowed. The non-arbitrariness of morality that categoricists mean, then, is that moral rules could not have been otherwise, whereas rules of golf, etiquette, baseball, chess, club rules, and all social conventions clearly could have been otherwise.

An exception may exist for moral norms offered as solutions to coordination problems. Which side of the road we ought to drive on is itself arbitrary, but learning this will not make us give up driving by the rules of the road. The categoricists may press this point, however, by noting that the normative imperative to drive on the same side of the road is not itself arbitrary. This point needs careful understanding. It is useful for traffic coordination if we all drive on the same side of the road, and once that is accepted, it is merely arbitrary which side of the road we choose, so long as we all choose the same side. So far, something similar may be said of golf: it is useful that everyone play by the same rules, and once this is conceded, then we can tolerate arbitrariness concerning the rules themselves, so long as we make it the case that everyone agrees (through an association and the publication of a rules book).

Categoricists would do better, then, by distinguishing solutions to coordination games from moral imperatives. "Drive on the right!" is a weak or Footian categorical, whereas "Don't murder!" is a strong, or Kantian categorical, and the difference between the two concerns the element of arbitrariness in the two commands. One might suggest the following demarcation: full arbitrariness entails hypotheticity, partial arbitrariness entails weak categoricity, and zero arbitrariness entails strong categoricity, or some such thing. Categoricists may even avail themselves of the Rawlsian-Scanlonian contractualist manoeuvre. For contractualists, any rational agent suitably situated with full knowledge would agree on (or select) the same moral rules. But no one can maintain that agents from behind a veil of ignorance, say, would come up with the rule about not grounding one's club in a hazard. Thus contractualists, closet categoricists after all, must accept a real distinction between social conventions and moral rules, and that this distinction has something to do with the strong categorical nature of morality. That agents are supposed to arrive at the same moral rules, the same principles of distributive justice, indicates that moral rules cannot be merely arbitrary, but must be a product of reason in some sense – a sense that the rules of golf and rules of the road are not.

Pointing out the arbitrariness in weak categoricity does not unravel the use of weak categoricity. Abandon the rules of the road at your peril. Moreover, we can easily admit that language has an arbitrary component in its origin without denying the use of language. If morality were arbitrary in this second sense, then it is hardly worth our concern. Of course, if we have some specific reservations about a specific rule, namely that it is counterproductive, then the arbitrary nature of its origin may finally propel us to abandon it for good. The same may be said for morality taken as a whole. If morality really were counterproductive, and its origin is arbitrary in the first place, we would have reason to abandon morality. By pointing out that the origin of the rule was not arbitrary, we can point out the reason for its inception, and that would be useful if for some reason we had forgotten that reason in our assessment. But the fact that there *was* a reason for the rule is not itself sufficient. If someone asks, "Why is *x* deemed impermissible?" it is a poor answer to explain how it originated. That discussion may have no bearing on its use now. That we are not allowed to drive faster than thirty kilometres an hour makes sense if we can point to a nearby elementary school, but not merely by appealing to its having once been a school zone, before they tore down the school, nor by appealing to the maximum pace of horse and buggies when they first came up with the speed limit. Of the moral rules we like, we must like them for a reason independent of their origin. We must like morality for what it does to us now. If so, the fact that this particular rule has an arbitrary origin is as irrelevant as the arbitrary meeting of our spouses.

What I have said above does not show that morality cannot be a strong categorical imperative. I have only shown that if morality were "merely" hypothetical (or weakly categorical), that would not be problematic. We should first appreciate the merit of a hypothetical morality before we can let go of the demand for categoricity. In other words, it is to no avail if we admit ideally rational agents suitably situated and with full knowledge would agree on a specific moral rule (like murder is wrong), in a way that we could never *ex ante* arrive at the same rules of golf or chess. It would not prove categoricity over hypotheticity. At least, it would not if the reason we would adopt one rule over another is to serve a particular interest that we happen to have, like keeping the peace, or living in a fair society, or understanding what heritable dispositions are needed for evolutionary fit. Each of these are predicated on circumstances that *themselves* might not have been, but for the contingent circumstances in which humans find themselves.

2.6. Intrinsic, Instrumental, and A Priori

But perhaps my highlighting the instrumental purpose of moral rules reveals another difference between strong and weak categoricity. Presumably our enjoyment of golf is itself the reason we stick to the game despite our learning of the arbitrariness of its rules. Admittedly we do not enjoy moral behaviour in quite the same way. To be sure, Aristotle spoke as if that were so. The prospect of enjoying moral behaviour for its own sake strikes me as psychologically false, but even if it were true, it does not exclude the possibility of indoctrination. Saying, "I do good because I am averse to the guilt I feel when I do bad," ignores the possibility that the guilt is due to an indoctrination independent of the wrongness of the act. Killing an unsuspecting Ojibway may have seemed a guilt-free act to a cowboy, for example. Masturbation may still seem a guilty pleasure to some. Saying, "I do good for its own intrinsic reward," apart from being psychologically displaced, is also theoretically problematic. If it is the doing good that I enjoy, does it matter whether what I take to be good really is good? And if it makes sense to even ask this question, then we need a criterion for determining whether an act is good independent of our intrinsic enjoyment of it. Conversely, if one enjoys doing good acts *because* they are good acts, the criterion to distinguish good acts must be understood independently of the enjoyment. Whereas golf, baseball, and chess are enjoyed intrinsically, not instrumentally.

Rules of the road, though arbitrary and hence only weakly categorical, are not observed for their intrinsic sake. A categoricist might use this to her advantage. One could argue that categoricity is presented as a third option to the intrinsic-instrumental dichotomy. We know that Kantian moral imperatives are not instrumental: that would be the calling card of hypothetical imperatives. Can categorical imperatives provide intrinsic reasons for acting? Then one would not be acting according to duty, unless it is the acting according to duty itself that carries with it some intrinsic benefit, perhaps the benefit of recognizing one is acting rightly. This is not the sort of intrinsic benefit connected to psychological well-being of which virtue theorists speak. Morality for Kantian categoricists is supposed to be a derivation of a priori reasoning, and a priori reasoning is not connected to happiness of either the intrinsic or instrumental variety. That is the domain of a posteriori propositions.

Thus a categoricist might maintain that, while weak categorical imperatives may be broken into intrinsic and instrumental groupings, strong categorical imperatives are something else: a priori. In this sense we could agree that categorical imperatives are not intrinsically rewarding without denying that they offer intrinsic reasons for action. This fits – albeit loosely – with Kant's tripartite distinction of hypothetical assertoric, hypothetical problematic, and categorical imperatives. Problematic imperatives, predicated on rules of skill, have instrumental use; assertoric imperatives, predicated on the happiness of the agent, have intrinsic use; and categorical imperatives are a priori. Pointing out a difference between the rules of the road and the rules of golf would not necessarily show that moral imperatives must be similar to one or the other if a third, a priori, distinction is lurking.

I assume, perhaps wrongly, that not all categoricists would wish to pin their hopes on the a priori nature of morality, but if my argument is accepted to this point, the hopes of categoricists rest on making the a priori clause stick. My dissatisfaction with the a priori move may be explained briefly. Basically the supreme categorical imperative cheats, and I do not mean this merely in the Hegelian sense that it is underdetermined. Kant's error is more egregious than that. The a priori move works only so long as we presuppose moral categories, not otherwise. But since the categorical imperative was to ground moral categories, a failure occurs. Take my maxim (my particular subjective volition) to be, "I want to steal this diamond necklace." Kant will say that I cannot logically satisfy my maxim in a world where taking diamond necklaces is universally permitted. After all, stealing entails the taking without permission, but if it is permitted, then I cannot steal. My maxim is not logically enactable should the maxim be universalized. That my maxim is not universalizable is a matter of a priori necessity. But this brilliant move is defeated should my original maxim not be morally loaded to begin with. It is not that I want to steal the diamond necklace. After all, I would not turn it down if it were offered to me as a gift. ("No, no, it isn't the necklace I want; it's the stealing of the necklace I want.") What I really desire is to have the diamond necklace. My maxim is, "I want to take this diamond necklace," or, if a maxim requires linking a means to an ends, "I take this diamond necklace because I like it." There is no a priori incoherence in universalizing this maxim. Stealing may well be an a priori no-no, but only to the extent that immorality is an a priori no-no. There is little to celebrate in that proposition. That my maxim of taking the diamond necklace passes the test, while my maxim of stealing the diamond necklace does not, shows the failure of pretending morality is an a priori necessity. If moral categoricity demands a priori reasoning, so much the worse for categoricity.

My above dismissal fails if we take Korsgaard's interpretation of Kant. Korsgaard suggests that Kant's test concerns practical incoherence, not logical incoherence.[23] The only constraint on our choice is that it has the form of a law. "[A]n autonomous will by its very nature must ... choose a law for itself."[24] The only restriction in choosing a law for itself is in terms of what the agent practically can will. That is, she must see whether her autonomous ends would be frustrated by universalizing this particular maxim. In this sense, however one phrases the maxim – with or without the moral terms – the practical implications will be the same. One cannot universalize the maxim of taking this necklace for it entails a law of people going about taking necklaces. This would have the effect of having the necklace one just took taken by someone else: thus the intention of having the necklace cannot be satisfied. The law cannot be rationally willed. All very nice, but as Mackie long recognized, this collapses the categorical imperative into a hypothetical imperative.[25] The practical impossibility of satisfying my maxim if it were universalized depends on the *contingent* notion that others will have similar maxims. There is no a priori truth to this. Its contradiction is not impossible to conceive. Its reasonableness is strictly on the basis of a posteriori reasoning. Moreover, the practical difficulty ensues because of particular preferences and interests I happen to have. That is, perhaps my desiring of the necklace is ephemeral at best, in which case I would not be practically foiled if someone else took "my" necklace after I have already lost interest in it. The practical impossibility is therefore neither a priori, nor independent of my interests in the way that a proper good will ought to be. My point here is that although I can agree with Korsgaard that the practical impossibility test is an excellent way of determining what is and what is not moral, it cannot be considered a categorical imperative in Kant's sense. It is contingent on the interests we and others happen to have and these interests are a posteriori. Any imperatives from these beginnings will be "merely" hypothetical.[26]

I have summarily ruled out morality's being either intrinsic or a priori. If I am right in both those cases, we are left with understanding morality as instrumental. Most people seem to accept the categorical nature of morality without thinking it depends on treating morality as a priori, but are they right about that? Is it conceivable that one can believe morality has instrumental purpose, but that purpose is nevertheless categorical? The imperative will always be contingent on satisfying the instrumental end, and absent that end, or the presence of an alternative means to that end, the strong categorical sense cannot be maintained. Accepting an instrumentalist account of morality cannot work in favour of categoricists. Thereby categoricists must defend the a priori understanding of morality, and the best of luck to them.

If morality is instrumental, it is enjoyed, not for its own sake, but only, or primarily, for what it can get us. But what is that? There are various opinions. For now, let us say moral rules tend to get us *x*. *X* may stand for well-being, social utility, fairness, considered self-interest, or evolutionary fit. None of this matters for my present purposes. If we enjoy *x*, and morality does indeed serve *x*, then we have a reason to adhere to moral rules. If we subsequently learn that the origin of those rules is arbitrary, what would that matter to us if they do indeed get us *x*? The intrinsic-extrinsic difference between sport and morality is irrelevant.

2.7. Semantics

Perhaps one last foray is available to categoricists: they may find an ally with semantically-minded golf aficionados. Certain semantically-attuned golf aficionados maintain that golf is played only one way – by the rules laid down by the governing bodies, and any substitution of any rule at any time constitutes a case of not playing golf. One is, perhaps, practising for golf, or playing a golf-like game – call it "flog"[27] – but it is not golf. If we desire to play golf, then we must follow the rules. Notice this is a hypothetical imperative. That we *must* follow the rules cannot be gainsaid by our imperative. We must follow the rules *if and only if* we desire to play golf. Many do not have that desire – especially when "golf" is so narrowly defined. The hypothetical formula leaves it open to allow persons to play flog to their heart's content. Is condemnation of these persons an appropriate response? Admittedly, we may condemn those who make avowals that they play golf narrowly conceived yet who do not in reality. Saying they follow the rules when they plainly do not is worthy of criticism, but our condemnation in such cases has more to do with dissimilitude than rules infractions. Compare those who openly confess that they do not play golf narrowly conceived and have no intention of ever doing so. They are quite happy playing flog. Condemnation of floggers for failing to honour the rules of golf seems out of line. Categoricists will claim that morality is different than golf in this respect. Condemnation of those who decide not to follow (approved) moral rules deserve our condemnation.

Admittedly, moral condemnation varies according to the severity of the social impact. Spitting in public may be immoral, but hardly warrants the same degree of condemnation as assault. Similarly, we could concede that those who play flog are "wrong," but it is a trivial sort of wrong. Although they ought to be playing golf, and not some bastardized version of golf, it will do no good to impose stroke penalties or disqualify players if the players are not in any sense competing, just as it is idle to threaten an F to a student not registered. Disqualification means nothing to floggers. They are simply engaged in a bit of recreation. Golfers may ostracize and belittle floggers, however, just as we may ostracize public spitters without going so far as imprisoning them. Through ridicule and disdain spitters may come to modify their behaviour. Likewise, through ridicule and disdain floggers may be heckled into becoming golfers. Morality works in the same way. Some moral infractions are too petty to add legal penalties, but social frowning is encouraged. Perhaps the same sort of low-decibel condemnation applies to floggers.

Fine, but to maintain the difference that categoricists require, let us note the existence of two types of non-golfers: those who play flog and those who do not golf at all. Even accepting some disdain for "non-real" as opposed to "real" golfers hardly justifies being upset at people who do not golf at all. Whereas, with morality, it is hardly an excuse to say, "Oh yes, I did violate that rule, but don't worry, I do not honour *any* moral rule." We might applaud them for their consistency, but we will still lock them up.

The semantic move fails, however. All that is being pointed out is that people hold others accountable, not that such accountability is categorically mandated. Foot agrees with Kant that morality cannot depend on self-interest, and so she accepts morality being categorical in that sense.[28] Her main objection is that, contra Kant, moral inclination depends on pretheoretical affections. "In my view we must start from the fact that some people do care about such things [the relief of suffering or the protection of the weak], and even devote their lives to them; they may therefore talk about what should be done presupposing such aims."[29] My concern is why such pretheoretical interests would become so widespread, and the answer will have something to do with interests in a way both Kant and Foot prematurely reject. (I shall return to this in 3.5.)

2.8. Kant's Conclusion

The preceding reflections are to leave the reader with the idea that there is no sense of strong categoricity that can apply to morality. Categorical rules are said to be universal, impartial, inescapable, unignorable, unconditional, non-contingent, necessary, absolute, good-in-themselves, and a priori. None of these conditions are uniquely met for morality, and the prospect of all of the conditions being jointly met is far fetched. Remarkably, Kant himself would appear to agree:

> Reason, therefore, restlessly seeks the unconditionally necessary and sees itself compelled to assume this without having any means of making such necessity conceivable; reason is happy enough if only it can find a concept which is compatible with this assumption.... [R]eason cannot render conceivable the absolute necessity of an unconditional practical law (such as the categorical imperative must be). Reason cannot be blamed for not being willing to explain this necessity by means of a condition, namely, by basing it on some underlying interest, because in that case the law would no longer be moral, i.e., a supreme law of

> freedom. And so even though we do not grasp the practical unconditioned necessity of the moral imperative, we do nevertheless grasp its inconceivability.[30]

This is an odd conclusion from a man who concedes that "[t]o posit a triangle, and yet to reject its three angles, is self-contradictory; but there is no contradiction in rejecting the triangle together with its three angles."[31] The same objection may be raised against Kant's odd conclusion to his *Groundwork*. Since we cannot understand morality as a categorical imperative without its collapsing into a hypothetical imperative, perhaps we should conclude that no such thing can satisfy the conditions of categoricity. If "reason" restlessly seeks the unconditionally necessary, and the unconditionally necessary is so far removed from human experience, it is time to curb such "reasoning."

2.9. Fictionalism?

If categoricity is false, is all of morality false? Or can a revisionist model of morality stand absent categoricity? Richard Joyce argues that categoricity is not merely an unnecessary appendage, it is the main ingredient. "Any system of values that leaves out categorical imperatives will lack the authority that we expect of morality, and any set of proscriptions failing to underwrite its authority simply does not count as a 'morality' at all."[32] He does not argue for abandoning moral discourse, however, but to adopt a fictive stance concerning morality, just as we persist in noting the beautiful "sunset," or to use Joyce's example, convincing ourselves we must do sit-ups. We know the "must" here is not categorical: it is only hypothetical. But we also know that should we allow ourselves to dwell on the hypotheticity of the "Must do sit-ups!" rule, we probably will not do sit-ups.[33]

I agree with Joyce that the categorical imperative is a myth. I disagree with Joyce in believing that therefore morality is a myth, since I do not view morality as a categorical imperative. Morality as a system of hypothetical imperatives suffices for social cohesion, and it is hypothetical strategies, not categorical strategies, that are successful in evolutionary game theory models. (I will discuss this more fully in Chapter 5.)

Certainly people speak as if morality is necessarily categorical (at least some of my rejection letters have been written by such people), but this alone does not decide the issue. Joyce and I know that is how they speak and Joyce and I know they are in error. The question is whether the error is fundamental to moral discourse or not. Would all morality crumble if we abandon categorical imperatives? Joyce thinks so, I do not. Consider the following statement: "A woman gives birth every seven seconds." When we agree with this, we are not lying to ourselves that there really exists a woman who spews out babies at a remarkable rate. Our agreement is not to render the utterance true of a particular woman, but as true of women as a class. Likewise, when we say, "I must do sit-ups," it is not as a categorical we agree to it, but as a hypothetical. We do deeply want something, to fend off for a while longer the belly bulge, and given that desire coupled with the belief that regular sit-ups can do that, we have a desire to do the sit-ups. Given the antecedent (about the belly bulge), we have an imperative to do sit-ups. We cannot say, as Joyce seems to think, that if we ignore the antecedent, it will appear to us as a categorical. That is trivially true. Moral utterances are the same. "I must not steal," may be an imperative conditional upon a line of reasoning that suggests not stealing leads to better success in society coupled with the real desire for success in one's society. It is not categorical. It is more a statistical claim. Generally speaking, people do better not stealing, and since you are a person, chances are you will do better not stealing. There need be no lie about this.

Importantly, Joyce recognizes this much. He argues, "If an agent is thinking only in terms of hypothetical imperatives – thinking that he ought to φ because doing so will, in the long run, satisfy his desire for x – then there is room for temptation to enter in the form of rationalization."[34] But if categoricity is endorsed as a fiction, this temptation remains. After all, "it cannot be argued that one *must* adopt the moral fiction – when the must is read as a categorical imperative."[35] To this objection, Joyce replies that a fictionalist will know why he is endorsing the fiction. "The whole point of the moral fictive stance is that it is a strategy for staving off inevitable human fallibilities in instrumental deliberation."[36] But this admits one may sabotage oneself to avoid the temptations without the fiction – for this bit of practical rationality is what prompts one to adopt the fictive stance. In other words, a motive independent of the fiction moves one to adopt the fiction, and so one is not moved by the fiction itself. Talk of fiction, then, is not needed. The appeal to do sit-ups is dependent solely on the hypothetical conditions moving me to make the seeming categorical utterance.

Joyce offers other arguments in favour of fictionalism. (1) When an error is found in a particular discourse, depending on the kind of error, one may either abandon the discourse entirely, like the discourse on witches and phlogiston, or modify the discourse to rid the error, like discourse on physics once we move from Euclidean to Einsteinian geometry. Alternatively, one may adopt a fictionalist stance, like "Must do sit-ups." Joyce thinks ethical discourse is more like the latter. To the suggestion that we merely revise our theories of morality to accommodate hypotheticity, rather than accept an error-theory about morality, Joyce argues that would be akin to merely revising one's theories about witches once we remove the bit about supernatural powers. His point is that the bit about supernatural powers is the *essential* bit of witch discourse, and removing that is to remove witch discourse entirely. Joyce's claim is that the same goes for morality: removing categoricity is to remove the *essential* bit of morality. There is nothing left to talk about once that is gone.[37]

> In the case of witches, for example, the whole point (one might say) of having a witch discourse was to refer to women with *supernatural* powers. To discover that no human has supernatural powers is to render the discourse pointless.... By comparison, the point of having a "motion discourse" was to refer to the change in position of objects in space over time. There was never a particular need to refer specifically to *absolute* motion. The fact that people thought of the motion as absolute was not a vital aspect of the discourse.... So the question we must ask is: "Is moral discourse more like talk of witches or more like talk of motion?" – and the answer is that it is more like talk of witches.[38]

The witch analogy is not apt. A revision is possible with morality much like a revision in our understanding of motion. One can be in error about morality in less fundamental ways – even if the view held in error is deemed fundamental by those holding the erroneous view. After all, many theists hold that God is fundamental to morality, but rejecting belief in God does not mean we must reject belief in morality. That those very theists would think so is not informative. In this sense, I offer a partial-error theory. Moral categories have use, and forming moral heuristics will also have use, but people's estimations of morality from those heuristics is where they go astray. They have overextended the heuristic. By highlighting the mechanics of how morality has use in terms of evolutionary fit, we highlight all that we are entitled to say about morality. Moral theory can be judged according to how well it accords with what evolutionary accounts entitle us to say.

(2) Without categoricity, no rights can be maintained. After all, rights claims entail obligations. If *A* has a negative right to *x*, everyone else has an obligation to refrain from taking *x* from *A*.[39] But an obligation is not necessarily categorical. I have an obligation to pay my membership, but this is only contingent on my wishing to remain a member. Of course, Joyce qualifies his worry: "If there are no inescapable moral obligations, for instance, then there will be no inviolable claim rights."[40] The trouble here is that terms like "inviolable" and "inescapable" are part of the categoricists' language, so cannot count against a system of hypothetical imperatives.

(3) In Joyce's appeal to fictionalism, he brings Hume in as an ally. Reason leads Hume to believe $\sim x$, while nature leads Hume to believe x. Joyce interprets this as Hume's maintaining the belief $\sim x$ while acting x.[41] But Hume's conclusion is not best read as $\sim x$, but that since $\sim x$ clashes with what nature demands (x), then the procedure that arrived at $\sim x$ is flawed, or at least limited. Analogously, if reason leads me to abandon morality, so much the worse for reason.

(4) Robert Nozick argues that if something has evolutionary fit, it is because of the way things are in the environment, i.e., something that is in fact true. That is, once you admit the usefulness of a trait, you admit the trait is useful because it is predicated on something that is true. Nozick's example is that if there is evolutionary fit to believe that two plus two equals four, this is so because two plus two actually equals four.[42] Likewise, if moral categories have fit, there is something true about moral categories, or true about how moral categories in fact have fit in the environment. Richard Joyce disagrees, complaining that Nozick's example cheats. Two plus two necessarily equals four, but we find no similar necessity in moral utterances.[43]

Joyce is right to complain about the math example that Nozick employs, but wrong to read so much into it. A contingent example will do for Nozick. An impala's speed has evolutionary fit for the impala because of the existence of certain predators in the impala's environment. We would not imagine the development of speed to that extent if there were no speedy predators. To say the development of speed would have come about (been useful) even if no fast predators existed in the impala's environment is to have a peculiar notion of evolutionary fit. Yet this is Joyce's contention: "Now imagine instead that the actual world contains no such requirements at all – nothing to make moral discourse true. In such a world humans will still be disposed to make these judgements (most generally to believe that categorical requirements exist), ... for natural selection will make it so."[44]

It is the usefulness of moral categories that makes the development of moral categories a successful trait. And the usefulness of them entails that they are not mere fiction. Rather, their use is predicated on something true about the environment in which moral category users live. Do not misunderstand me here. Admitting the usefulness of moral categories does not entail that our beliefs about those moral categories are true. That a conventional norm is founded on something true in the environment does not by itself prove the conventional norm itself is true. The norm may be comprised of unnecessary bits that piggy-back on mis-described useful bits. Malaria, for example, was first thought to be caused by bad air (*mal aria*), and the prevention for bad air was to shut the window. Shutting the window does not prevent bad air: if anything it prevents the flow of fresh air. What shutting the window did do, however, was to keep the infected mosquitoes out, the real cause of malaria. So here the convention of shutting the window worked, but for reasons different than the cultural beliefs.[45] Despite the error, the success of the cultural phenotype of shutting the window *was* based on something true about the world. Likewise, many of our moral beliefs may be false – categoricity is a prime example – but it does not follow that the success of moral behaviour is due to our false beliefs, let alone due to a specific false belief – like our belief in categoricity. Joyce takes the key element of moral beliefs to be the categorical imperative and if so, I would agree with him that moral beliefs are false. But given the evolutionary benefit of moral action (and presumably the belief that generates that action), a better conclusion is to insist that moral judgements have been incorrectly identified as categorical.

(5) For Joyce, the alternative to an error theory is to merely stipulate a hypothetical definition of morality. "[S]ince long-term selfish judgments turn out to favour cooperation, promise-keeping, etc., such judgments just *are* moral judgments (or, at least, such judgments are a proper subset of judgments of 'enlightened self-interest.')."[46] Joyce rejects the stipulative move for the following reason: "But this strategy unravels when we consider the instances where there can be no serious case made that an agent's ends *are* served by cooperation, coupled with the observation that moral discourse may still adamantly condemn him."[47] A hypotheticist may say either of the following: (1) If his ends are not met, then we want to say he has no moral compunction, or (2) Yes, but although he is treated as immoral by the rabble, *that* is where the mistake lies, not in the first formulation. In other words, all Joyce is pointing out is that the practice of morality is not consistent with a hypothetical definition of morality, but since the motive for coming up with a hypothetical definition is that the current practice of morality is indefensible and incoherent, Joyce's objection is entirely idle. Besides, the evolutionary picture of morality that Joyce accepts teaches us not to defend particular moral actions by straightforward appeals to rational self-interest. That fails as much as the categorical story (a point I will make good in Chapter 4).

When we say we *define* morality in a certain way, the definition is to point out the common features of most moral utterances, so that we may be more consistent in our application. Let us say that someone holds an act to be immoral because it has the features ABCD, leaving aside what these features or relations are. We can criticize the moral statement in one of two ways: either by appealing to facts or by appealing to the definition itself. To appeal to the facts, we can point out that the particular act in question is not a case of ABCD. Alternatively, we can say, "Although it is a case of ABCD, you are in error in defining moral infractions as the occurrence of ABCD." But admitting an error concerning the conjunctive proposition A and B and C and D is not to show that some of those components, say ABD, are wrong. Perhaps it is only C that we wish to challenge. In this case, C stands for categoricity.

Notes

[1] "[T]here remains nothing but the universality of a law as such with which the maxim of the action should conform" (Immanuel Kant, *Grounding for the Metaphysics of Morals*, James W. Ellington (trans.), (Indianapolis, Ind.: Hackett Publishing Co. 1986) Ak. 421. [Page numbers to Kant refer to Paul Mezner's text (Berlin, 1911) as it appears in Vol. IV of the Königliche Preussische Akademie der Wissenschaften edition of Kant's works (Ak.).]

[2] "Now all imperatives command either hypothetically or categorically. The former respect the practical necessity of a possible action as a means for attaining something else that one wants (or may possibly want). The categorical imperative would be one which represented an action as objectively necessary in itself, without reference to another end" (Kant, Ak. 414). "Finally, there is one imperative which immediately commands a certain conduct without having as its condition any other purpose to be attained by it. This imperative is categorical. It is not concerned with the matter of the action and its intended result, but rather with the form of the action and the principle from which it follows; what is essentially good in the action consists in the mental disposition, let the consequences be what they may. This imperative may be called that of morality" (Kant, Ak. 416).

[3] "The reason for this is that whatever is necessary merely in order to attain some arbitrary purpose can be regarded as in itself contingent, and the precept can always be ignored once the purpose is abandoned. Contrariwise, an unconditional command does not leave free the will to choose the opposite at its own liking. Consequently, only such a command carries with it that necessity which is demanded from a law" (Kant, Ak. 420).

[4] "Now, if the action would be good merely as a means to something else, so is the imperative hypothetical. But if the action is represented as good in itself, and hence as necessary in a will which of itself conforms to reason as the principle of the will, then the imperative is categorical" (Kant, Ak. 414). "[T]he categorical imperative alone purports to be a practical law, while all the others [rules of skills and counsels of prudence] may be called principles of the will but not laws" (Kant, Ak. 420).

[5] Kant, Ak. 416.

[6] Kant, Ak. 420.

[7] Kant, Ak. 416.

[8] We may deduce that only morality is a categorical imperative from the following: "[T]he categorical imperative alone purports to be a practical law" (Kant, Ak. 420), "... the categorical imperative, or law of morality ..." (Kant, Ak. 420), and, "Hence there is only one categorical imperative, and it is this: Act only according to that maxim whereby you can at the same time will that it should become a universal law" (Kant, Ak. 421).

[9] Phillipa Foot, "Morality as a System of Hypothetical Imperatives," *Philosophical Review* 71 (1972): 305–16. Foot's objection to Kant is that morality is not alone in offering categorical utterances. "The conclusion we should draw is that moral judgments have no better claim to be categorical imperatives than do statements about matters of etiquette" (Foot, 312).

[10] In competitive tournaments it is expected that everyone knows and follows the rules. But golf is more generally played by amateurs who compete only with themselves, or the course, not with other players. My commentary has to do with this latter group.

[11] John Searle, "How to Derive an 'Ought' from an 'Is'," *Philosophical Review* 73 (1964): 43–58.

[12] See note 3.

[13] Any moral theory prescribes positive and negative obligations. Anything not covered by the obligatory is by default morally permissible. That is, every action by humans will be under the auspices of a moral theory. Whether morally permissible acts can be further divided into prerogatory and supererogatory we can leave aside.

[14] Richard Joyce, *The Myth of Morality* (Cambridge: Cambridge University Press, 2001), 36.

[15] A strong categorical must be an all things considered ought, but not the reverse.

[16] Some may protest that brushing one's teeth is predicated on health, and that health is not mere convention. Bloomfield, for example, makes exactly this point (Paul Bloomfield, *Moral Reality* (Oxford: Oxford University Press, 2001)). Fair enough, but we still have the problem in moving from descriptive facts of things which cause unhealth to the normative claims concerning what one ought to do about that. For more on this, return to my discussion at the end of 1.1.

[17] Becker argues that the difference between morality and etiquette does not concern the inescapability of the rules, but the scope of the rules. Moral considerations, he argues, are overriding considerations in ways that rules of etiquette or baseball are not. And "when we have 'considered everything' in so far as we feel bound to follow reason at all, we will feel bound unconditionally by the ought (the moral ought) we have reached." Lawrence E. Becker, "The Finality of Moral Judgments: A Reply to Mrs. Foot," *The Philosophical Review* 82/3 (1973): 369.

[18] One might, instead, defend a rule (a new rule, especially) by invoking the integrity of the game. This is the approach the USGA took to limiting the COR efficiency in drivers. In that sense, the purpose of the rules is to facilitate play in some sense.

[19] Foot believes rules of etiquette are also defeated by discovering their arbitrariness (Foot, 312). If this were true, etiquette would have dissolved long ago. Perhaps it has.

[20] An exception may exist for moral norms offered as solutions to coordination problems. I return to this below.

[21] For these cases, see Hilary Putnam, *The Collapse of the Fact/Value Dichotomy and Other Essays* (Cambridge, Mass.: Harvard University Press, 2002), 109. He concludes, "we need no better ground for treating 'value judgements' as capable of truth and falsity than the fact that we can and do treat them as capable of warranted assertability and warranted deniability" (110).

[22] Putnam, *The Collapse of the Fact/Value*, 121.

[23] Christine M. Korsgaard, "Kant's Formula of Universal Law," *Pacific Philosophical Quarterly* 66 (1985): 24–47.

[24] Christine M. Korsgaard, *The Sources of Normativity* (Cambridge: Cambridge University Press, 1998), 220.

[25] J. L. Mackie, *Ethics: Inventing Right and Wrong* (Hammondsworth: Penguin Books, 1977), 27–30.

[26] One might object that I am looking for a Kantian a priori justification in the wrong place – in the application of a categorical imperative and not in the formula of the universal law itself. (Thanks to Thaddeus Metz for raising this.) A Kantian can readily admit that a posteriori considerations play a role in determining a particular application of the universal law without being committed to supposing a posteriori considerations play any role in determining the universal law itself. My intent is to cast doubt on deriving moral rules from a priori reasoning, not merely to point out the obvious: that application of moral rules requires a posteriori reasoning. Teasing apart the application of a rule from the derivation of the rule, I think, makes my case all the clearer, however. For one could easily grant that given a particular rule, a priori reasoning mandates certain applications. For if it is immoral to do *x*, and *A xs*, then *A* is immoral, and this we can understand by a priori reasoning. But determining that it is immoral to *x* on a priori grounds itself is where the trouble lies. The root of the problem rests with Kant's believing (or my disbelieving) a priori judgements can also be synthetic. Is "$2+3=5$" synthetic? Frege did not think so. (Frege reduced the truths of arithmetic to analytic truths of logic. Gottlob Frege, *The Foundations of Arithmetic*, J. L. Austin (trans.), (Evanston, Ill.: Northwestern University Press, [1884] 1953), §§1–4, 12–17, 55–91.) Is the judgement, "The shortest distance between two points is a straight line," a priori? Not under Einsteinian geometry. When Kant's own examples fail, the onus lands on the shoulders of its defenders.

[27] Thanks to Mark Yaniszewski.

[28] Foot, *Philosophical Review*, 312.

[29] Foot, *Philosophical Review*, nt 15.
[30] Kant, Ak. 463.
[31] Immanuel Kant, *Critique of Pure Reason*, Norman Kemp Smith (trans.), (New York: St. Martin's Press, 1929), [I, Pt. II., Div. II, bk. II, ch. III, sec. 4], 502.
[32] Joyce, *The Myth of Morality*, 176–7.
[33] Joyce, *The Myth of Morality*, 215.
[34] Joyce, *The Myth of Morality*, 213.
[35] Joyce, *The Myth of Morality*, 222.
[36] Joyce, *The Myth of Morality*, 223.
[37] Joyce, *The Myth of Morality*, 157.
[38] Joyce, *The Myth of Morality*, 96.
[39] Joyce, *The Myth of Morality*, 175–6.
[40] Joyce, *The Myth of Morality*, 175.
[41] Joyce, *The Myth of Morality*, 190–91.
[42] Robert Nozick, *Philosophical Investigations* (Oxford: Clarendon Press, 1981), 342–8.
[43] Joyce, *The Myth of Morality*, 164–5.
[44] Joyce, *The Myth of Morality*, 163.
[45] Joe Heinrich, Wulf Albers, Robert Boyd, Gerd Gigerenzer, Kevin McCabe, Axel Ockenfels, and Peyton Young, "What is the Role of Culture in Bounded Rationality?" in Gerd Gigerenzer and R. Selten (eds.) *Bounded Rationality: The Adaptive Toolbox* (Cambridge, Mass.: MIT Press, 2002), 355.
[46] Joyce, *The Myth of Morality*, 222.
[47] Joyce, *The Myth of Morality*, 222.

Chapter 3

SELF-INTEREST

3.1. The Problem

If morality is a system of hypothetical imperatives, what is the antecedent condition that morality serves? Hobbes's Foole demands that morality must serve his self-interest if morality is to motivate him.[1] Agreeing with the Foole on that point does not solve the problem, however, since there are many ways in which morality may be said to be predicated on self-interest, not all of them compatible. If morality is to serve self-interest, how is this argument to be made?

Self-interest is both vague and ambiguous. As Bentham notes, "Interest is one of those words, which not having any superior genus, cannot in the ordinary way be defined."[2] When speaking of self-interest, one may think of a strict psychological egoism, so that acts of altruism are done, not for the sake of others, but for the sake of the self. Such a thesis has little hope of defence, but no self-interested account of morality need be interpreted so narrowly. Interests for contractarians are simply whatever desires the agent happens to hold, and this should include any moral sentiment he wishes to act on; including benevolence. As Bernard Gert rightly decrees, "for Hobbes, it is simply a matter of definition that all voluntary acts are done in order to satisfy our desires. But... he does not deny that we can desire good for another."[3]

A more serious rift in understanding interests concerns whether interests are objective things or subjective things, whether they are externally ascribable to agents or only internally ascribable to agents.[4] If my interest is external to me, something is in my interest whether I endorse that interest or not. I would have a reason to *x* whatever my avowed interests are. Objective accounts come in two varieties that are often not clearly distinguished. If we see someone who smokes, drinks, eats fatty foods, and is lethargic, we are induced to believe she is acting against her interest in health. Whether she *ought* to have such an interest and whether she *does* have such an interest are two different matters. On a normative interpretation of the objective account, we will say, "Given her behaviour, she has no interest in her health, but she ought to." Good grounds for why she ought to have such an interest will have an external locus that carries normative force for the reasonable.[5] A descriptive interpretation of the objective account of self-interest, on the other hand, claims she really does have an interest in her health but for some reason is ignoring it, perhaps due to *akrasia*.[6] In the first case, we can say she has no interest in the very thing in which she ought to have an interest. If so, we can give up saying someone has an interest in something she shows no interest in – a misuse of language. She fails to have a worthwhile interest. She does not desire the desirable. In the second, she ought to conform her beliefs to match interests she really has, albeit unrecognized by her.[7] This is the meaning Butler had in mind when he utters, "There is a manifest negligence in men of their real happiness or interest in the present world, when the interest is inconsistent with a present gratification . . . thus they are often unjust to themselves as to others."[8] On both accounts, her avowed claims of self-interest are irrelevant to what is really in her self-interest.

Neither interpretation of objective self-interest will be of any comfort to the Foole.[9] By the brute fact of his why-be-moral question, the Foole understands self-interest subjectively. The things the Foole takes an interest in are what the Foole does not want to give up by behaving morally. We cannot placate the Foole by telling him that his giving up those interests is compensated by the satisfaction of interests he does not admit to having. Of course, we can give a reason without presuming this will provide motivation.[10] If motivation and reason are separated like this, the Foole is concerned with a motivating reason. Providing an unmotivating reason, then, is not an answer to the Foole's question.

Demanding motivating reasons is to demand internal or subjective reasons, reasons the agent herself can endorse. A subjective rendering of self-interest tells us that what we desire we call "good,"[11] but this seems flawed in at least two distinct ways.[12] The first is that there is no guarantee that subjectivists have not simply got the ordering backward. It is not that I think friendship good because I prefer it, or that I think deception bad because I prefer not to be deceived, but that it is because friendship is good that I prefer it, and it is because deception is bad that I wish to avoid it. We prefer that which is good.[13] That subjectivists get the ordering backward is made more evident when we ask *why* we desire something. It is odd to suppose preferences can be completely arbitrary. Why would I prefer x unless there is something inherent in x that makes it so that I prefer it?

My preferences must track *something*, one would imagine. Desirability, thereby, must reside in the object and not in the mere desiring.

The second problem is that we can be mistaken about our interests. Our experiencing regret is evidence of such a fact. We constantly do things we think in our interest which later turn out to be bad for us. How can we be mistaken about what we take to be good if good is defined as that which we take it to be? More fitting is to say that what you desire you *believe* to be good, though it is not necessarily so. It is a common belief that some things are good for a person whether or not she prefers them. Getting an inoculation may stand as an illustration.

A corollary of this second problem concerns the evident inability of subjectivist accounts to accommodate reflection about our interests. Subjectively avowed interests are varied. Some of them conflict. In order to avoid regret – not an irrational goal – we must decide among competing interests. More, we must decide whether the satisfaction of a present preference is better to us than the satisfaction of another equally pressing interest. To do this, we must weigh our subjective desires against some background other than what we subjectively desire. A purely internal standard appears to be indistinguishable from no standard at all.[14] There is no criterion an agent can appeal to without it being a criterion she endorses at the time. For example, while drinking, one prefers drinking more to avoiding the hangover. While being hung over, the preference ordering flips.[15] Which conflicting passion best represents her subjectively defined interest cannot be settled merely by appealing to what the agent decides at the time. The problem of clashing interests results precisely by focusing too narrowly on the tug of present time-slice desires.[16] Therefore, a subjective account of what is good cannot easily account for reflection about desires.[17]

What follows is my reply. In 3.2 I shall respond to the reverse-order complaint. In 3.3 I shall explain how subjective relativists and internalists can accommodate mistaken interests. One might agree with me on these issues, however, without agreeing with me that we can forge a link between subjective interests and morality. Something else has to be said, and this typically concerns an appeal to considered subjective preferences. The notion of a considered preference has problems, however, which I illustrate in 3.4. My stance on how morality and self-interest link will be given an overview in 3.5, and further elaborated in Chapters 4 and 5.

3.2. Reverse-Order Objections: Three Replies

1. A preliminary difficulty with the reverse-order objection follows from accepting the mistaken preference objection: they cannot both be right – or at least there is a tension in holding both. In the mistaken preference

objection, it is claimed a person may think a thing good that is not. In the reverse-order objection, it is claimed a person's desires are attracted to things that are inherently good. Goodness is what drives the attraction. If so, I cannot be attracted to that which is not objectively good. Yet, this is precisely the assumption embedded in the mistaken interests problem. Either we deny that people can desire wrongly, or we permit criticism of desires. The fact that we are attracted to things bad for us speaks against the reverse-order objection. How we come to desire something may be criticized (for example, children may prefer name brands over cheaper brands due merely to peer pressure and manipulative advertising), but to deny that they do desire the thing seems specious.[18]

The claim that everyone desires the good is false if "good" is to be interpreted as an objective, moral good. Whatever you take to be an objective moral good, not everyone desires that. Saying they *ought* to is a confession that they do not in fact. On the other hand, the claim is uninterestingly true if we treat "good" in a subjective relative amoral sense. The thing that drives desires are what people take to be good. We need not presume a homogeneity among what people take to be good. The underlying similarity is merely the structure of desiring itself, not the content of that desiring. In this sense, "People desire the good," is a tautology. Given this distinction between the two senses of "good," the reverse-order objection would have to explain how it could be possible for someone not to desire the good if the reason we desire the good is because of its goodness. In the subjective sense, the ordering is irrelevant, given that it is a tautology. In the objective sense, the statement is empirically false.

One might think reverse-order objectors need say only that *sometimes* something is objectively good, and this will motivate the desiring. They need not suggest the hopelessly false proposition that this is so *all* the time. Nor need they suggest the absurd doctrine that the objective goodness of the property is what causes the desiring in agents properly so attuned (like a certain auditory signal striking a functioning ear will cause the hearer to hear a noise). Clearly it is true that what we take to be good we hope is really good, and sometimes our hopes are met. But this happy contingency does not undermine the subjective-relative amoral sense of good. For an agent need not call x good if, despite what others tell her about the goodness of x, she does not desire x. It is not a good *for her*. The reverse-order objection requires a case where the goodness is recognized in the agent prior to the agent's desiring it. Such a case will not be found without merely presupposing the objective goodness of things independent of an agent's desiring. On the other hand, if we admit that sometimes A desires x because x is really good and that A may sometimes be mistaken in what A takes to be good, then we admit that *A* may desire *x* because *A* believes *x* to be good – which is different from suggesting that *x really is* good. All we could say is that desires are predicated on what individuals take to be good – even if sometimes what they take to be good really is so. Recognition of mistaken preferences thereby undermines the reverse-order objection. Being moved by mistaken preferences highlights that all we can say for sure about what moves us is that we are moved toward objects we subjectively deem good, not which really are good. As Blackburn notes, "[t]he fact that there is a cannonball on the cushion explains why it is sagging in the middle. The fact that kindness is good explains no such kind of thing."[19]

The fact of acquired tastes might count in favour of the reverse-order objection. With acquired tastes, we have had to learn what was good about a product and we are sometimes glad we did. But this does not show that some things ought to be desired. It may rather show that we are malleable by nature and can come to desire almost anything once acclimatized.[20]

There are two further replies to the reverse-order objection. The first requires a mental state account of preference satisfaction, and the second charges the reverse-order objection of begging the question.

2. Take the obvious fact that people differ in their preferences. This ought to be puzzling if preferences simply track something inherent in the objects of desire or aversion. To accommodate heterogenous interests, there must be something in the person, and not just the thing, that drives preferences. One plausible suggestion points toward a mental state account of preference fulfilment rather than a success theory of preference fulfilment. Under a success theory, aims are satisfied if and only if they have been satisfied in fact, whether or not the agent is aware of it.[21] We make this distinction since it is possible that my aims have not in fact been satisfied; I may erroneously believe they have been. On a mental state account, what counts for me as preference satisfaction is my belief or feelings at the phenomenological level that my preferences or aims have been satisfied, whether or not they have been in fact. On a mental state account, when I say *x* is good, it is so because it produces in me a mental state that I endorse. Typically mental state accounts are summarily dismissed,[22] but I hope here to resurrect them.

Let us say that I desire that my enemy be destroyed and that my children thrive. Presumably this means that I would *really* like my enemy be destroyed and my children to thrive, as opposed to my falsely believing my enemy is destroyed and my children thrive. That is, it appears my state of mind is incidental to my desires. If it comes to a choice, I prefer {my enemy destroyed, my thinking they are not destroyed} to {my enemy not destroyed, my thinking they are destroyed} and {my children thriving, my thinking they do not thrive} to {my children not thriving, my thinking they do thrive}. Granted, best of all is {my enemy destroyed, my knowing about it} and {my children thriving, my knowing about it}, whereas worst of all is {knowing that my enemy is not destroyed} and {knowing that my children do not thrive}.[23]

Such preference ordering is typical.[24] It undermines the claim that happiness or self-interest is purely a subjective, or internalist affair. We are not after states of mind, but states of the world. Epictetus's picture is the reverse. Happiness is merely the satisfaction of desires, but what those desires are is incidental. If you are unsatisfied, you may either try to alter the world to have it conform with your desires, or alter your desires to fit the state of the world. The former, according to Epictetus, is foolish since the world is not prone to obey your commands.[25] Admittedly, Epictetus may have exaggerated both the difficulty in changing the world to meet one's desires and the ease of changing one's desires to fit the world, but his advice is completely insane if we reject mental state accounts. If Epictetus's advice has any merit, and surely it does, then the simple rejection of the mental state account cannot do.

Take for example our children's thriving. What does it mean to me that my children thrive? Is it that they have wealth and health enough? Would my desire for their thriving be satisfied if despite their health and wealth they are themselves discontented? Would I be content in my desire for their thriving if despite lack of wealth and health, they are content with their lives? And if they desire that I am happy while my happiness is contingent on their desires being met, we are in a loop. Such a loop is not solved by appeals to the states of the world. The point of these questions is to suggest that even if I am concerned for states of the world, the states of the world in these cases are the mental states of others.[26] Of course I may be deceived about their mental states and this is not what I would wish. On a mental state account, my being deceived is no different from my not being deceived up to the point that the deception is discovered. The introduction of a deceiving pill or experience machine is supposed to successfully remove this latter condition and our reluctance to take this pill is supposed to count against mental state accounts.[27] Our hesitation, however, may well be due to our doubting the grand claims of the thought experiment. As Brandt remarks, if we know we are simply a brain in a vat, it is unlikely we can at the same time enjoy experiencing otherwise.[28] To willingly agree to enter the pleasure machine or ingest the deceiving pill is to recognize one is being deceived – a recognition that is incongruous with what the argument is intended to show. The thought experiment is ineluctably contaminated.[29]

Perhaps the problem people have with mental state accounts can be addressed by delineating wishes from desires, as Aristotle suggested.[30] Let us call "things for which we can aim" as "things which are in our power of acquiring." If so, wishes are for things toward which we cannot aim, while desires are for things toward which we can aim. Perhaps I have a desire to play the violin. Perhaps, as well, I am unsure of my ear. My tendency to believe I am improving then, cannot be based on my own judgement, of which I have little confidence. Therefore, my assessment of my improvement is positively correlated with external evidence. Perhaps people no longer cover their ears. They sit and listen longer. They make compliments. They inform me that I am improving. Likewise, my belief that I am failing is also contingent on the external world: on others. All that I can aim at is self-satisfaction with my violin playing. I can wish others appreciate my playing, but that is beyond my full control. Since it is beyond my full control, it cannot count as a desire. It can only count as a wish. Coupled with Epictetus's notion that only mental states are in our full power to control, desires will be restricted to mental states.[31]

A different objection to mental state accounts may occur in the following manner. We may experience pleasurable mental states while wishing we did not. Being aroused inappropriately may stand as an example. The arousal in itself is pleasurable, but it is conceivable that I would prefer not to be aroused at this time or in this circumstance. This shows that our desires can trump mere mental states, and therefore mental state accounts do not present the full picture. This follows, however, only so long as we do not attribute any mental states to the second-order preference to not being aroused in this circumstance. To discount second-order mental states, however, according to Parfit, collapses into an absurdity.

> Suppose that you are in great pain. You now have a very strong desire not to be in the state that you are in. On our revised theory, a desire does not count if you would prefer not to have this desire. This must apply to your intense desire not to be in the state you are in. You would prefer not to have this desire. If you did not dislike the state you are in, it would not be painful. Since our revised theory does not count desires that you would prefer not to have, it implies, absurdly, that it cannot be bad for you to be in great pain.[32]

The reverse-order objection has more merit if we take a success theory stance rather than a mental state account. But a success theory is thought superior to a mental state account only if we take the reverse-order objection to be telling.

3. An alternative reply to the reverse-order objection is open to the subjectivist that is independent of reliance on a mental state account. To claim

A desires *x* "because it is good" is by itself unrevealing. "Because it is good" is a slippery notion. Since one is free to understand "it is good" in subjective terms, the difference in ordering is of no consequence. "I desire *x* because I deem it good" is pretty much the same as claiming "good is that which I desire." The only way the ordering can make a difference is if what we mean by "it is good" is somehow independent of my so deeming it, a stance moral realists endorse wholeheartedly.[33] It is precisely that proposal that is rejected by the internalist, however. As a result, an external rendering of the notion of "good" is not shown by this criterion: it is rather presupposed.

Can an objectivist avoid the question-begging charge by making a distinction between epistemology and metaphysics? For example, saying one prefers that which is good (rather than calling good that which one prefers) is not metaphysically asserting that there is an objective goodness. Rather it makes the epistemic claim that objective goodness figures into the best explanation of some desires.[34] In this sense, it is not an illicit appeal to objective desires: merely an inference to a best explanation for a convergence of preferences across persons on certain things.[35] As far as tastes go, for example, people tend to prefer chocolate to chalk. Surely it is not misguided to presume such a convergence in preferences has something to do with the actual properties of chocolate and chalk, and not merely that people *happen* to prefer one over the other as if it were merely a matter of choice. Of course preferences and desires are not quite the same, and preferences based on direct neurochemical stimulus is a different sort of thing than desires for following the path of moral goodness in interactions with others. So we need a connection between the chalk-chocolate example to desires for moral goodness as similar kinds of good abductions. And all that the subjective relativist need highlight is that as far as moral goodness goes, no such connection can be found. For the thing that makes it sensible in the chalk-chocolate example is empirically driven. We can point to an objective property in chocolate. We can point to an objective property in our tongues and neural mechanisms (in "normal" Homo sapiens). There is no analogous empirical property to point to in considerations of the good. To make the abduction from chocolate over chalk preferences to an objective moral goodness, we would have to hypothesize a similar neurochemical reaction in normally functioning Homo sapiens brains to moral actions. Such a hypothesis can hardly count as the "best explanation." As Mackie might say, that is the *worst* explanation. Moreover, when we morally appraise someone's action, we could not care less whether they actually satisfied some objective interest they had – we appraise what they do based on *our* interests – and it little matters to us whether our interests are objective or subjective, so long as they are our interests.

Admittedly the chalk-chocolate analogy does highlight one thing: desiring may be triggered by objective facts in the world. If what I have argued in Chapter 1 still holds, however, morality is not an objective fact in the world in any comparable sense, and so moral goodness – the thing we are ultimately interested in – cannot be understood to be driving the desiring the way something in chocolate and normal Homo sapiens taste buds drives the desiring for chocolate. For that matter, even though we may say that an interest in chocolate is based on some objective fact, we would not say that *therefore* the interest in chocolate is good, or that *therefore* her desiring chocolate is good, let alone that *therefore* she who has an interest in chocolate is good. We are inclined to say she who has an interest in helping others is good, that the interest to help others is good, and that an interest in helping others is good. We would say such things even if the interest in helping others is not found to be due to any objective fact in the world.

Part of the above discussion rehashes one of the debates between Mill and Moore. For Mill, "the sole evidence ... that anything is desirable, is that people do actually desire it."[36] Desire, Mill claimed, is analogous to sense perception. That which is visible or audible is that which is in fact capable of being envisioned or heard. Thereby, that which is desirable ought to be simply that which is capable of being desired. Moore rejected the analogy between desire and sense perception. That which is desirable is not simply that which is able to be desired by someone, but is instead that which *ought* to be desired, or that which is *good* to be desired.[37] Although people tend to see the same things, people do not desire the same things. Moore's objection here is right. Since only those desires of mine that I endorse are desirable in the appropriate sense, it is not true that what is desirable is simply what is desired. The extensional sets of the desirable is smaller than the extensional set of the desired. Still, this does not warrant the dismissal of a subjective account of the desirable, since it is possible that anything desirable to an agent must also be desired by that agent. This reading is not ruled out by anything Moore says above. If the desirable must be a subset of what an agent desires, the desirable cannot be given a criterion independent of the agent's own desires. As such, "proper desires" can be understood in terms relating to the agent's own reasons, not as desire rankings that an external observer holds, and Moore's critique leaves this subjectivist understanding open.

But can a subjectivist account of self-interest distinguish between proper desires and mere desires? The second objection doubts this and to that I now turn.

3.3. Mistaken Preferences

We have many desires. Satisfying one may be at the expense of another. We need a way to decide which preferences warrant our approval, and which do not.[38] To seek to avoid regret is to admit that we can be mistaken about our interests. Mistakes, properly categorized, do not count against a subjective account, however. Mistakes amenable to subjectivism come in four varieties:[39] (1) Mistaken beliefs (mistakes due to unavailable or inaccurate information and mistakes due to faulty processing of information); (2) Type Mistakes (or category mistakes); (3) Behavioural disabilities (mistakes affecting motivation); and (4) Mistakes due to changed values or desires. A putative fifth category of mistake concerns failure to desire the properly desirable: a mistake that is not amenable to subjectivism, obviously. What I shall argue is that all types of mistakes that objectivists point to can be classified under any of the first four categories: those consistent with subjectivism.

1. Mistaken Beliefs. Regret may be due to mistaken beliefs. I desired x. I believed A would secure for me x. I was wrong. My being wrong about A does not entail that I was wrong about x. Such mistakes concern the chosen *means* to my ends, and do not concern the ends themselves. Perhaps experience will help me choose better means in future. Of mistaken beliefs, there are two kinds: (a) those due to unavailable or faulty information, and (b) those due to faulty processing of available information. Although such mistakes undermine one's success at achieving one's ends, they are not the sort of mistake that undermines a subjectivist account of interests. There are yet other mistakes about information we make concerning our desires, preferences, and interests.

2. Type Mistakes. Category or Type mistakes appear less like mistakes about the *means* to one's ends, but a mistake about the very ends themselves. Type mistakes occur when one confuses local for global desires. Global desires are those most in accord with the agent's most basic desires.[40] Other desires are worthy of pursuit so long as they lead to the satisfaction of her basic desires, or are in fact constitutive of her more basic desires, or at least do not contravene her most basic desires. Global desires are more basic to personal well-being than local desires. More than local desires, global desires capture the person we want to be, or the kind of person we discover ourselves being when most happy. Global desires are more constitutive of who we take ourselves to be.[41] By contrast, local desires are more easily discardable; they are specific to context and psychological state.[42]

Most proffered objective lists will tend to be the sorts of things global goals are seeking to attain: autonomy, liberty, accomplishment, love, friendship, health, etc. For subjectivists, items on objective lists will be desired not for some metaphysical communion with Truth, but rather the more mundane explanation that the list is derived from a descriptive account of what people do tend to globally desire. But individuals will rank these various items in ways more conducive to their subjective preferences when items on the list conflict – accomplishment over health, for example, or contentment over accomplishment.[43] Criticism can occur, however. Once we admit that individuals have both local and global preferences, and we rate global preferences as carrying more weight in relation to our overall well-being or interest, subjective-relativists are free to criticize an agent's choosing to satisfy local preferences at the expense of her subjectively avowed global preferences. The endorsement of local goals are thereby derivative of our subjectively defined set of global goals. Local goals that further approved global goals are ipso facto approved. Interests that do not conflict with global goals are approved by default. Interests that conflict with approved global goals count as mistaken interests. So long as I have a global goal that I endorse and know will be curtailed by this local distraction, abiding by the local desire anyway permits criticism.[44] If I desire to count grass and this desire interferes with my global desires, we can say grass counting is not in my interest.[45]

The type mistake occurs by confusing local for global goals. Bank robbery, for example, is pursued not with the intent of acquiring wealth, presumably, but with the intent of acquiring that which wealth provides. It is not a risk-free enterprise, however, and failing to accurately assess the probabilities of being caught and winding up in prison may be deemed a mistake of reasoning. Unless the intent of the bank robbery was the thrill component itself,[46] the mistake is in defining the bank robber's interest as "robbing the bank." Presumably that was deemed instrumental to his interest, and not itself the interest.[47] Thus, although we can complain about the methodology employed to secure his interest, we cannot complain about his higher-order interest (admittedly materialistic). In assessment of our acts, we must not fail to look both at the global interests our acts were trying to achieve, as well as at those global interests our acts frustrate. We may thus judge the efficacy of acts by relating to the basic global interests they were intending to satisfy. One of the more successful rehabilitation therapies for convicted criminals follows this model.[48] Type mistakes are open to criticism within a subjective framework.

3. Behavioural Impediments. Despite clear desires and good information about how to achieve those desires, we sometimes have a behavioural impediment to act. A standard behavioural block is the phenomenon of *akrasia*. *Akrasia* is commonly described as having an insufficient will to act on one's preferences, but this obscures matters. *Akrasia* involves a weighing of the cost of the means with the benefit of the ends. Let us say A prefers x to $\sim x$. To get x, A needs to do y. A prefers $\sim y$ to y. Whether A prefers x to $\sim y$, or $\sim y$ to x determines whether she can be charged of *akrasia*. Should she all things considered prefer $\sim y$ to x, we cannot charge her of *akrasia*. Likewise, should the problem lie, not in the person, but in the incommensurability of x and y, then again we cannot charge her with *akrasia*. *Akrasia* can only apply in the case that she clearly prefers x to $\sim y$, yet nevertheless fails to choose y. It is difficult to believe this is as common a phenomenon as is supposed,[49] but should it occur, criticism is warranted. Our criticism, note, is independent of an objective standard: it is based precisely on the preference ordering of the weak-willed agent: that is, our criticism can be understood in terms the agent herself endorses.

Other behavioural impediments exist. Obsessive compulsive behaviours and phobias may inhibit action aimed at fulfilling one's desires. Likewise, delusions, schizophrenia, bipolar disorders and the like may inhibit accurate assessment of available information. Psychological help at remedying these problems is not impermissible on a subjective relativist position, but the focus of this help has to respect an autonomous agent's right to refuse treatment. Also, the intervention must focus on removing the obstruction, not forcing upon the subject what choice she would make absent her behavioural impediment. What choice of action is best for me in an idealized context is not clearly the best in my actual situation. I know how I would act absent vertigo, but it hardly follows that the best course of action for me is to act as if I did not suffer from vertigo. Absent vertigo, I might successfully make my leap across a chasm. But given my vertigo, it is less clear I can muster the requisite confidence to successfully leap, and to be told the best course of action for me is to leap notwithstanding this impediment is, quite frankly, a mad theory. I may very well prefer not to suffer from vertigo; but to *ignore* this constraint in deciding my best course of action can hardly be claimed a self-interested act. Even if the Foole suffers behavioural impediments, therefore, it is not yet clear that morality is straightforwardly in his interest. Of course, psychological derangement is not typically ascribed to the Foole.[50]

4. Altered Desires. Regret may be de due to a change in values or desires. I desired x, *A* secured for me x, but now I no longer desire x. My regret about acting on *A* may not then be due to my mistake about *A*, but my having since changed my desire for x. My regret about acting on a value or desire I now no longer hold hardly challenges my belief that I should act on values or desires I hold. I still endorse that belief. My regret is that the desire or value I held is not the one I hold now.[51] The problem against subjective self-interest accounts, then, must go deeper than pointing out the mere fact of regret. The issue is what preventative measures are open to us to avoid regret due to later altering our desires; to make as sure as possible that the desires we have now are those we will continue to hold. To do this, objectivists argue that we must reflect not merely on the best means to achieve our present desires, but also on the very desires themselves. To consider what desires one would likely keep over time, one must move beyond a simple catalogue of what one's present desires are, to somehow assess those desires in relation to how well they will stand the test of time. On what basis can such an assessment take place if we are to retain a subjectivist account of interests?

One solution to the reflection problem is to move away from actual subjective desire fulfilment theories to counterfactual subjective desire fulfilment theories. That is, if we can appeal to the desires the agent would endorse under epistemically idealized positions, we can avoid the charge that subjectivist accounts cannot accommodate reflection.[52] Such reflection about *information* is permissible on the subjective-relative model of self-interest. A reassessment of one's global preference itself on the basis of an idealized stance, however, smacks of aristocracy. The more idealized position this counterfactual account demands, the less possible it is for real people to be moved by it. It certainly will not assuage the Foole. We are moved by social and psychological pressures and biases, and to know what act rational agents would choose absent these constraints may motivate change, but there is nothing necessary in such reflection. That we would choose *A* over *B* in a counterfactual domain is not obvious. That we *can* be moved by pure reason absent social and psychological constraints is even less obvious. If I have an ineradicable irrational passion, a rational act for me ought to accommodate it, not dismiss it. As with the case of vertigo, what is best for an agent must take into account the agent's personality, biases, foibles, dispositions, beliefs, fears, motives; not discount them. Decisions made under idealized positions are decisions made within a different context from actual decisions and do not necessarily apply to the situation at hand.[53]

Perhaps one of the best challenges to the subjective relativist position concerning self-interests is made by appeals to psychotherapy. Both Brandt and Bloomfield point out that we often are not the best judges of ourselves, and it sometimes takes others to tell us what is best for us.[54] If this were not true, there could be no niche for therapists. This truism need not undermine the internalists' position, though, so long as we keep in mind that the observers take their stands not from some idealized notion of what a good person is supposed to be, but from paying closer attention to this particular agent and all her various desires and values and ambitions.[55] There is nothing impossible in a therapist or friend observing, "Gee, all things considered, maybe you should keep counting grass."

Summary. The standard complaint against a subjective rendering of self-interest is the recognition that we can be mistaken about our interests. That we can be mistaken about our interests need not undermine a subjective account of self-interest, however. We can be mistaken about our interests in relation to other interests that we hold. The mistake will reside in the relative weight the agent assigns the interests in the light of other facts and desires she endorses. Admittedly, more needs to be said. My concern here, however, is with something else. Even if we accept a purely internalist account of interests, how can we convince the Foole that he is wrong in thinking his subjective interests are best satisfied by immoral acts. Is it necessarily the case that the Foole's belief structure that connects satisfaction of his interests within the constraints of dynamic social interaction is internally faulty?[56] Internalist ethicists normally commit themselves to showing that so long as the Foole's interests avoids (1) mistaken beliefs, (2) type mistakes, (3) behavioural disabilities, and (4) temporary desires, he should discover that morality will always be in his interest. Can *this* claim be seriously maintained? In the following section I shall argue that it cannot. In Section 3.5 I shall argue that this bit of bad news need not undermine an internalist account of interests, nor a hypothetical account of morality.

3.4. Considered Preferences

When answering the Foole, no one proposes a purely subjectivist reading of interests. No one pretends that morality serves the Foole's subjective desires, whatever they may be. To argue that one thing, morality, serves *everyone's* self-interest in this strict subjective sense is far-fetched. The subjectively interpreted interests of a rapist, of a murderer, of a thief, are hardly well served by the moral restrictions of their pursuits. When Hobbes, for example, speaks about subjective self-interest, he means to highlight a subset of subjective self-interest – what we will call, after Gauthier, "considered self-interest." These are interests which stand after the Laws of Nature have their way. The worry about the claim that morality is in the considered interests of every individual is two-fold. On the one hand, we may wonder whether the criteria for consideredness slips in moral constraints. On the other hand, if we are careful not to slip in any moral presuppositions in our understanding of considered interests, we can show that morality does not serves everyone's self-interests. I shall argue that it is impossible to avoid both worries. To satisfy one is to commit the other.

The contractarian argument for linking morality to self-interest goes as follows: Morality serves the purpose of better enabling individuals with disparate aspirations to cohabit peacefully. As individuals prosper within a moral realm, as compared to an amoral realm, it is in each individual's self-interest to live in a moral domain. As the moral domain cannot flourish without general obedience to the moral dictates that the system advocates, it is, indirectly, in each individual's self-interest to obey those moral dictums. My reason for adopting the moral system is different than my reason for abiding by the system. Although I may abide by the system of morality merely out of fear of punishment, I can accept such a system so long as the rules protect me in ways I deem, in sum, profitable for me. In other words, punishment cannot be arbitrary. Institutionalized punishment must meet my self-interest, largely construed.

Richard Joyce, with whom I agree on other points, gets this wrong. He argues that when we morally condemn a person's actions, we are not complaining that she harmed *herself*, but has harmed others, and *therefore* morality cannot be a mere case of self-interest.[57] Such thinking conflates what morality is (harming others without their consent) with the reasons agents have to endorse morality. Although there is no self-interest in being punished, there are self-interested reasons for endorsing a system that has accompanying punishments.

Imagine, for example, driving while drunk and as a result crashing and being seriously injured. In what sense, we might ask, do we mean the act was motivated by "self-interest"? Butler appears to be right when he says some acts are "really" in our self-interest, while other acts are not.[58] It is this bit of common sense contractarians endorse in claiming that morality, if it is to further individual interest, must further the individual's *considered* interests rather than her actual interests. An individual's considered interests are those interests that are *rational* for that individual to pursue, and not merely whatever interests an individual currently happens to believe she holds.

Considered preference theory holds that one has a reason to *x* if and only if one would be motivated to *x* after a process of fully informed correct deliberation. The process of fully informed correct deliberation is not to be taken as an external, or objective reason. An external claim, to use Williams's terms, is one that is applied to the reasoner regardless of her desires. Although it is common for people to believe in external reasons, Williams makes the case that all such talk is false. For Williams, reasons must be able to explain an agent's action, and to explain action, the reasons must relate internally to that agent, i.e., reasons must be motivating, or potentially motivating to that agent.[59]

Yet the ability to distinguish internal from external reasons when we are speaking about "correct deliberation" and "potentially motivating" is problematically vague. Jean Hampton argues that "correct deliberation" entails an external reason itself. If an agent fails to deliberate correctly, then her internal reason does not count. More importantly, we may say she has a reason to deliberate correctly whether she knows it or not, and *this* normative advice is external to her.[60] Joyce complains that Hampton's objection introduces one too many imperatives. It is not that an agent ought to deliberate correctly no matter her interests. After all, she may have no interests she wants to pursue at present. Rather, if deliberating correctly is imperative to her getting what she does want, *x*, then she ought to deliberate correctly. The "deliberate correctly" counts purely as an instrument to achieve her internal desires. It is not meant as, nor should it be interpreted as, an imperative external to whatever her desires are.[61] But there is still the problem about "potentially motivating." To describe counterfactuals as internal reasons stretches things. Let us assume, with Hobbes, that whatever people seek they call good.[62] If they seek to satisfy their unconsidered preferences, it cannot be the case that what they consider good for them is the rational act. They will be unmotivated to pursue it. The contractarian claim is that since *x* better enables people to pursue their disparate goods, people rationally, if not in reality, will see *x* as a good. However true that may be, there will be a conflict for a person *A* who does not see the good of *x* in this light. If good is defined in terms of self-interest alone, then *x* will not be good for *A*. If it is not good for *A*, it is irrational for him to pursue it (assuming what is rational is to pursue one's notion of the good). At this level, then, contractarians advocate a contradiction. *X* for *A* is both good and not good. The worry is whether reliance on "considered self-interest" as a grounding for morality is simply like saying: "Morality is based on self-interest, so long as your interests are moral."

An out for contractarians is an appeal to hypothetical agreements. After all, contractarians do not (normally) hold that morality is determined by actual agreements. Rather, morality as an institution is justified by what rational people *would* consent to in the original position or in the state of nature. Asking what people would actually agree to places far too much weight on their current situations, situations that are very different for different people. Actual choices, in other words, are largely determined by the situations in which the agents are embedded. If our concern is whether these are the situations they *ought* to be in, we must be able to step outside that situation – otherwise the evaluation will be flawed. Just as we cannot justify a speed limit merely on the grounds that exceeding it will get you a ticket, we cannot properly judge our current social structures if our judgement is biased by the habits and conditioning of those very social structures.[63] At the meta-level of preference evaluation, all our preferences are necessarily hypothetical. The expectation that these hypothetical preferences, preferences that reside outside the domain in which agents find themselves, are "considered" is thereby not unreasonable. That choices made in this hypothetical context differ from choices embedded within the actual situations of the world is neither surprising nor problematic. The different senses of good (good for the actual agent and good for the contractarian) show no contradiction; merely different levels of discourse.

Unfortunately, this manoeuvre is too facile. For Rawls, we decide a moral structure once and for all from behind a veil of ignorance.[64] The difficulty is that we have no evident motivation to step behind that veil when we know the outcome will be counter to the material preferences of our actual situations (at least for the haves of the world). To say it is the fair thing to do for the haves to make moral decisions from behind the veil is to beg the question. We know it is moral for the haves to *x*; what we wanted to know is whether *x*-ing is in the haves' self-interest. With Gauthier, on the other hand, the motivational problem is supposed to be avoided: the considered preferences are understood as a specific set of our actual preferences. We must see that cooperation will benefit us in our actual lives. And this can only make sense if we assess this benefit from the standpoint of our actual preferences; not our hypothetical ones. Our considered preferences are to be a subset of our actual preferences, albeit, perhaps, somewhat buried beneath the various unconsidered preferences that we also hold. To be motivating, considered preferences cannot be preferences of a different domain than actual preferences. Gauthier clings tightly to internalism.

If moral claims are to be supported by interests persons have, then it cannot be a moral recommendation to have different interests than they do have. The existence of hypothetically desirable states will not satisfy the Foole, for he can easily admit this at the outset. If he were suitably motivated to be moral, he would indeed have motivation to be moral.[65] Telling Mary she would find utility from helping should she have an intrinsic preference to help is comparable to telling someone who does not like mushrooms that she should like mushrooms on the grounds that she would then be able to derive satisfaction from those mushrooms. Such advice is uninteresting. The Foole's question is why ought he be so motivated when he does not have the antecedent conditions. If we are to seriously address the Foole's question, we must do so in terms of the subjective interests he endorses, backed by the arsenal of criticism open to internalists. The Foole is entitled to ask, "Why should I give weight *now* to aims which are not mine *now*,"[66] and internalists must think this question sensible. To answer the Foole, moral action must be shown to further interests the Foole now endorses. The Foole claims this task is impossible. Talk of considered preferences merely kicks up dust, for there is nothing in an internalist's account that prevents us from ascribing to the Foole considered preferences. Perhaps he desires love and acceptance (a considered preference) but wrongly thinks that riches is the best way to secure love and acceptance, and that defection in prisoners' dilemmas is the best way to secure riches. In such a case, we can point out that riches is not a good way to secure love and respect, and in any event, honing the disposition to defect is certainly not going to achieve love and acceptance. But is the global desire of love and acceptance something that someone with considered preferences must have? Nothing in the internalist's tool kit can demand that. Besides, the Foole may want love and acceptance only from a select few, those toward whom he agrees not to defect, but this cannot convince him not to defect against others.

The main worry is that no internalist account of considered preferences can mandate that the Foole is not acting on his considered preferences. Part of this follows from the vagueness in distinguishing considered from unconsidered preferences. A preference is considered so long as, after sufficient reflection, one would not alter that preference. "[O]ne may have reason to act contrary to one's actual occurrent preferences given that, were one adequately to reflect, one would change those preferences."[67] But what is the process of reflecting on one's preferences? What is the mechanism of considering one's preferences? Under Gauthier's program, it cannot be preference – for that is what is being assessed – nor can it be rationality – for what is rational is defined *once* the preference is considered.[68]

If we do insist on "considered" preferences, that is an admission that rationality is *more* than merely preference maximizing; it must also be involved, as James Fishkin noted, in preference *ordering*. "The criteria for considered preferences specify nothing about the appropriate conditions for preference formation."[69] Since Gauthier conceives value as subjective, the person who determines whether a given preference is considered to the requisite degree must be the individual agent herself. If while intoxicated, the Foole decides it is a considered preference to drive home, then it appears to count as a considered preference, whatever the tragic outcome. Please note that I am not arguing against the efficacy of considered preferences over unconsidered preferences. Simply, the subjective experience of one's preferences *at the time* of acting is such that one believes the preference is sufficiently considered. Otherwise, one would not act upon it. Admittedly an observer of the Foole's recklessness may have a different opinion, just as the Foole himself may as he recuperates in the hospital, but this is beside the point. As Baier noted, "There does not seem to be any genuine empirical way to determine when someone has sufficient experience or has reflected sufficiently to see whether or not he has a fully, or even sufficiently, considered concrete preference."[70] No one other than the agent herself can determine whether her own preference is sufficiently considered. Gauthier, importantly, recognized this. "For if it be agreed that values are subjective, then there is no ground for appeal beyond what a person acknowledges." And if this is true, then we cannot accept the qualification that Gauthier adds, "given that she reflected sufficiently and is fully experienced."[71] It is this qualification that causes problems. As outsiders, we cannot circumscribe her actions based on *our* ideals of considered preferences. If she does not make any further reflection, then it is precisely because she assumes (or does not think otherwise) that she *has* sufficiently reflected on the matter. Given this, so long as she acts on a preference, it is symptomatic that she deems that preference to be a sufficiently considered one. In thinking that it is considered enough, the phenomenological account of that preference is that it *is* considered. By this account, every preference of hers would be considered (to her) – no matter how "unconsidered" they may seem to someone else or even to the same person at a later time. The only way, then, to conceive of "considered preferences" as a meaningful distinction is to base the criteria of consideredness on external factors – a violation of the internalist's credo.

I am not denying that it is often in our interests to resist acting on unconsidered preferences. This is not a book on well-being. Maximizing well-being is not the province of morality for narrowers (see 1.3). The question I am addressing is whether morality is necessarily in one's interests, and if one begins to answer that with a qualification of what counts as a proper interest, one's answer is "No!" So the emphasis on considered preferences already misses the mark. Keeping to a strict internalist's diet, a subjective criterion for what is considered can provide no criterion at all, since, flatly, every act would be deemed "considered." When one acts, that is the choice one thought right at that moment given the circumstances. To maintain that morality is grounded on considered preferences is to introduce a standard that is independent of the agent's preferences. It introduces a test where the criterion for what counts as considered is found to be external to the individual's subjective ranking.[72] Grounding ethics on considered preferences conceives morality externally, not internally; in terms of standards, not preferences.[73] Appealing to standards that override preferences is to abrogate the raison d'etre of internalist ethics; namely to offer an explanation for why morality satisfies subjective preferences better than amorality or immorality. To satisfy this motivation, it is imperative that the preferences and interests appealed to are *actual* and not merely those preferences and interests one *ought* to hold, or what actual agents *would* hold if they were ideally rational. By redefining what we mean by "individual preferences" to only those preferences that are sufficiently considered is an appeal to what preferences people *should* hold, and not to what preferences they hold in fact. If we appeal to standards that determine what interests individuals *should* possess, then it follows that a particular moral rule is really in a person's self-interest to abide (not merely to have others abide) *whether or not that person thinks so*. This, needless to say, repudiates the contractarian claim of linking morality to individual preference. If we judge proper preferences according to objective tests or societal standards, it is no longer individual preference that is maximized by obeying moral rules.

Summary. The criterion for what is a considered preference collapses to either an objective standard that is external to the individual agent, or to a purely subjective standard whereby the agent herself determines whether her preferences are considered. If it is objective or standard-bound, then we have lost the motivational force for an agent to be "rational" when her preferences dictate some alternative choice. If morality is to be based on rationality, and rationality is pre-determined by an objective criterion for what preferences are *worth* pursuing, we have introduced a normative concept into the definition of preferences, something utterly at odds with Hobbes's original intent. This would be fatal to the preference-based theorist's goal of linking morality to individual self-interest. On the other hand, if the determination of considered preferences is left wholly up to the individual, then we must admit that a considered preference can be none other than a preference held presently by that individual – whether or not it will sustain sufficient reflection. If this is true, then the introduction of "considered" preferences is wholly idle and we have, in fact, no different an account of rationality than that which furthers one's preferences. Introducing considered preferences as a criterion for rational self-interest, then, is either fatal or idle. Gauthier insists "that the preferences be held in a considered way, and this is not a matter of what considerations the person cares to give, but what would in fact survive adequate experience and sufficient reflection" – which he admits to be "vague standards."[74] They are certainly vague standards. Making them any more explicit will either abandon the internalist's stance, or reveal that the Foole need not be suffering from unconsidered preferences. As it turns out, we can do without the introduction of considered preferences.

3.5. Preferences, Schmeferences

When speaking of preferences in relation to contractarianism, one would expect two questions: what are preferences? and what role do they play in contractarian theory? The second presupposes some agreement on the first. I would like to argue here that this latter assumption, with one exception to be explained below (3.5.3), is false. We could not care less what preferences are. Nor should we care what their ontogeny is. Whether we are dealing with interests, preferences, values, desires, or needs should not matter. Whether we think preferences are subjective, or objective, or both, whether we think they are primarily cognitive states or emotive states, whether they may be autonomous or non-autonomous, considered or unconsidered, none of that should affect moral theory. Morality's role is not to serve preferences, but to solve interpersonal preference conflicts. As a solver of interpersonal preference conflicts, morality must be non-partisan to preferences themselves. This is the key to answering the Foole.

3.5.1 The Conventionality of Preferences

Preferences and interests are culturally malleable. That we have an interest in food is pretty universal, but the preferences we have for the sort of foods we choose to satisfy the basic caloric need ranges widely. Lobster meat among the early Acadians was considered the arachnid of the sea and was tolerated only as fertilizer. Later, some of the more desperate poor ate it. Children were mortified to find lobster meat sandwiches packed in their school lunches, and preferred to throw it away rather than suffer the abuse from their peers.[75] Now lobster is a delicacy – even for Acadians. The taste of lobster meat has not changed, nor the taste buds of Acadians. The cultural attitude toward lobster meat has changed. Few in Canada would think of eating horse, though horse meat is lower in cholesterol and fat than most other red meats. The Mapuche of Chile, however, frequently eat horse meat (and rely on horse travel, as well). Among the Machiguenga of the Peruvian Amazon, nobody eats snake meat. Anyone eating snake is thought to "throw up blood and die."[76] Just down the river from this Machiguenga village is a Mestizo village inhabited by immigrants from the Andean highlands who eat snake with no such adverse effects. Canadians side with the Machiguengas when it comes to snake meat, but not with the Machiguengas' penchant to snack on the fat-rich larvae found under decaying logs. And while we consume lemon and grapefruit, they squish up their noses at that as we would when they offer us a serving of slugs.[77]

Meanwhile, North American toddlers will happily eat bloodworms, solidified chunks of bacon fat, cigarette butts, house flies, and the like. From where does the later disgust to these things emanate? Is it experience? No noticeable trauma is associated with eating many of the things we grow averse to. Is it a change in taste buds on the tongue? No physical evidence suggests such a thing, although the brain is still in development, and there may be neural alterations unexplored. But that would not explain differences across cultures. More likely, children mimic the disgust overtly displayed in the culturally adept folk around them. Who counts as the culturally adept? Years of carefully training my daughter not to be squeamish with insects was quickly annulled once she met her peers in public school. We fashion ourselves after those we identify as prestigious. Prestige will be interpreted in terms of success, and success is interpreted in terms of the group with which we identify.[78] I am not talking about this being the best education; I am talking about how we are in fact educated.

If morality is based on enhancing our preferences, and preferences are cultural indoctrinations, it does not follow that morality is culturally relative. The role of morality is independent of preferences.

If preferences are culturally relative, the only normative appeal concerning preferences is conformity, not rationality. We, like most (perhaps all) animals, are mimickers. What you identify as your preference may not be yours at all. It may be an inculcation, a cultural meme. And even if it were replaced, what would it be replaced by but another cultural meme? Rationalists might complain: "But should your present cultural memes have a net cost for you, it is in your interest to upgrade your preferences." True enough, but what is a "reason" but itself a cultural meme? What you call a "net cost" is itself a cultural meme.

In what sense can it be in my interest to alter my preferences when interests are defined relative to the preferences I have now? That it would be in my interest to buy mushrooms if I like mushrooms is not in question. That it is in my interest to buy mushrooms even if I do not like mushrooms is. It is not that we cannot change our preferences – it is that we cannot defend a grand normativity about preferences. (We can defend a weak normativity predicated on conformity. That is, you ought to conform your preferences to your dominant group, but this "ought" is a built-in evolutionary strategy, not a rational decision. Rationally, we balk at the idea.)

The connection of preferences to moral normative advice is tangential at best. But even this minimal connection is wildly misunderstood. Consider the prisoner's dilemma (PD), where mutual defection is the rational solution, while mutual cooperation is both the moral and the preferred solution. (For a full account of the PD, see 4.2.) One might say one does better if one changes one's preferences from defection to cooperation, but this would be to misspeak. Defection and cooperation are not preferences. They are strategies. One's preferences remain: try to get the highest net utility possible. The normative advice concerns the *means*, not the end. Preferences and interests are not found in the means.

Differences in preferences, therefore, should be irrelevant to moral theory. That you prefer outcome x and I prefer outcome $\sim x$ should not matter when it concerns which outcome is moral. Otherwise morality moves, or is rational to, only those who happen to have the right preferences. To say, "Now you ought to have the right preferences" means only that you ought to be moral. And of course that is a moral ought. To say morality serves your preferences only so long as you have the right preferences is an admission morality does not serve preferences.

What would be needed is a separate argument that having the right preferences is a good thing without appealing to morality. It is a good thing for you independently of morality. Now normally when we say x is a good thing for you we mean x serves your interests: it is advantageous to you in some tractable sense. If the "tractable sense" offered does not relate to you (i.e., is merely external), then this cannot count as being in your interest as defined above. The general must be tractable in terms of the particular. Your interests will, in some concrete sense, mandate your preferences. If no preferences are satisfied in pursuit of a putative interest, we say the putative interest is not yours. We do not say there is a real interest here which your preferences fail to match. In any event, since what we are asking you to do is to alter your preferences, we cannot appeal to your preferences. But what else moves you?

Or does this argument move too fast? If I ask you to invest some money, my appeal is a greater return of money. We do not say I cannot appeal to your preference for money if I am asking you to invest some money. It is precisely your preference for money I am appealing to. Admittedly, I must appeal to something other than your preference for your *current* money. Analogously, I can appeal to you to alter your current preferences by the promise of a greater return in future preference satisfaction.

Two problems with the investment analogy exist. One is that money is interchangeable, preferences are not. If I have five dollars earmarked for a baseball card, it does not matter to me whether it is *that* five dollars or another five dollars so long as I get the baseball card. But our preferences are not interchangeable like this. In a world where technology investments are reaping higher returns than pork rind investments, I would do well to switch from pork rinds to technology. To say, "Yes, but I prefer to invest in pork rinds rather than technology," is weird. It reveals that I am missing the point. I am playing some other game, not the investment game. Now consider the following. In a world where mushrooms are plentiful and cheap and broccoli rare and expensive, I would increase my overall preference satisfaction by preferring mushrooms and disdaining broccoli. To say, "Yes, but I don't like mushrooms, I prefer broccoli," is *not at all weird*. What would be weird here is to say I ought to change my preference ordering in order to increase my overall utility. It puts utility ahead of preferences. Utility is a measure of my extant preference satisfaction, not an ideal toward which I should shape my preferences. Of course, if broccoli is not merely hard to come by, but completely unavailable, crying about one's preference for broccoli will be tiresome. In such cases, the inability to satisfy one's preference is a reason to wean oneself of that preference. We return to Epictetus's counsel. The problem of unappeased preferences is not the problem the Foole faces, however. He claims to achieve his considered preferences through defection; he is not crying for the unachievable.

The second problem with the investment analogy is that it does not capture the PD structure. I am not investing preferences when I opt to cooperate. The risk is that my investment will lose: I will have a net loss. This will happen if you renege. Of course, all investments have this fear, but this is not the risk most investors speak about. Rather, despite your partners's best efforts, countervailing events may interfere. The fad for the Atkins diet, say, may decline. You may lose your investments despite due process. In the case of PDs, the risk is counting on due process in the first place.

3.5.2 Game Theory

Classical game theory is to show one's interests/preferences are furthered by adopting conditionally cooperative strategies in PDs and other games. As far as game theory analysis goes, we do not care what your preferences are, so long as cooperation is suboptimal in terms of straightforward (or modular) rationality (otherwise there's no problem to solve). We are concerned about the relation of preferences, not preferences per se, just as logicians are concerned about the relation of the propositions, not the propositions themselves.

When conflicts over preferences arise among equally rational bargainers, the best strategy is to cooperate: to find a compromise. The "best strategy" is that which secures more of the agent's preferences. But what if our preferences are different than that described? Then you would not be in the conflict. But what if I have a preference to defect? Preferences to defect and preferences to cooperate are not preferences. They are strategies concerning how best to secure whatever preferences you have. One of these strategies is better than the other. We define "better" in terms of how many of your highest ranked preferences are most fulfilled. This is not a redundant statement. You have a number of preferences. Some of these matter more to you than others. Strategies that tend to attain more of your most preferred preferences are better than strategies than attain fewer, everything else being equal. But any preference may be partly filled, wholly filled, or not fulfilled at all. For that matter, preferences may be insufficiently or excessively filled. My preference for potato chips is not satisfied by having merely one potato chip. I find myself preferring to have zero potato chips than only a few. But nor is the preference more fulfilled with each chip gained. There is a point of over-satiation, even ignoring the cholesterol count. There is an optimal level of preference satisfaction for each preference. In terms of bargaining with others, then, there is much room for negotiation, both in terms of the number of preferences, the personal ranking of preferences, and the amount each preference is fulfilled.

Defection and cooperation, then, are not themselves preferences. In a PD, it is not that I prefer to defect. Defection is not a preference – it is a strategy. It is a good strategy for the agent only so long as it yields a higher payoff in terms of preference satisfaction than any other available strategy. Game theoretic results show that in iterated and in correlated PDs, this condition is false. Conditional (or mixed) strategies do better.

A critic may object thusly: "But we prefer the results of defection to the results of cooperation. Given a reverse ordinal ranking of the PD, we prefer 4 to 3 and 2 to 1, and since defection gives us either 4 or 2 while cooperation gives us 3 or 1, we prefer defection to cooperation. Who is this Murray to say otherwise?" Such thinking double counts, though. The reverse ordinal ranking places a numerical value on one's preferences. We do not say we prefer a certain ordinal ranking. The ordinal ranking is a measure of our preferences. The utiles are simply a marker of my preferences, not my preferences themselves. I have a preference for a certain state of affairs occurring over a different state of affairs occurring, and the chances of that state of affairs occurring under strategic negotiation is increased, not decreased, when I can adopt the strategy to conditionally cooperate.

To say to persons in a conflict over preferences, "Look, you guys will each do better adopting a disposition prone to cooperate with each other in terms of overall preference satisfaction," should be a true statement *no matter what preferences one has*. Let me clarify: it may be a false statement, but if it is to have any punch at all, its truth must be independent of what one's preferences are. The solution to preference conflict is the domain of morality. Morality is not something to which preferences must conform. Morality is not a competing concern (as if I have my preference, but I also have to do what is moral). Morality just *is* the best individual strategy for conflict resolution.[79] Notice that a strategy is different from an action. A good strategy need not entail an optimal move understood within the parameters of a given action, any more than a good bet guarantees success on a given occasion. Morality is a *means* to preference satisfaction given long-term strategic negotiations.[80] So in *this* sense, morality serves one's preferences. Morality concerns how best to fulfil your preference when that preference is contingent on the outcome of strategic interaction. When talking about morality, we do not care about what preferences are, where they come from, whether they are autonomous or nonautonomous, whether they are considered or unconsidered. When talking about well-being, that is another matter. Well-being entails the satisfaction of certain preferences. Since well-being has much to do with one's fit in society, those certain preferences should be approved by the group to which you identify. Which group with which you come to identify is shaped by one's parental and peer influences. In Rawlsian terms, a lottery. So the normativity involved in talk of well-being, even, is to be understood in no grand sense.

3.5.3 Non-Tuism

Concerning the above, a caveat is needed. So far, my thesis is this: *Resolution of interpersonally conflicting preferences is the domain of morality. Resolution of intrapersonal preferences is the domain of theories of well-being. Morality's role is not to serve preferences, but to solve interpersonal preference conflicts. As a solver of interpersonal preference conflicts, it must be non-partisan to preferences themselves. Whatever your preferences are, you do better cooperating.* This avoids the messy talk of considered preferences and internalism. But this basic position can be true only so long as we rule out tuistic preferences.[81] If we allow tuistic preferences, then we can have preferences concerning the satisfaction of others' preferences. If my preference is to have your preferences thwarted – whatever your preferences are – then the moral solution will not speak to me. Rather than say, "Ok, morality doesn't serve everyone's preferences," I want to say, "Look, that goal is not properly speaking a preference at all." Instead, tuistic preferences are metapreferences, just as Frankfurt's secondary preferences are metapreferences. They are preferences about preferences. My secondary preferences concern my primary preferences. My tuistic preferences concern someone else's (primary or secondary) preferences. In this sense, to think of tuistic preferences (good or bad) as a preference *simpliciter* is to commit a category mistake.

Given that morality's role is to solve interpersonal conflicts over preferences, good tuism is not an issue for morality to solve: there is no conflict (except in the case of two individuals equally moved by good tuism, but then the difficulty is a looping problem[82]). Serving one's bad tuistic preferences will normally conflict with moral advice.[83] To ask which higher-order preference serves my primary preference better, the answer is the one that satisfies most of most of your preferences (not a redundancy there). A preference to thwart other people's primary preferences will also thwart one's own primary preferences, not merely in those cases where the primary preferences are shared. Perhaps I like broccoli, but not mushrooms, while you like mushrooms, but not broccoli. A perfect cooperative venture is available where I get the broccoli and you get the mushrooms, and both of our preferences are maximized. But if, apart from my primary preference for broccoli and against mushrooms, I have the secondary preference to thwart your wishes, one strategy is to demand I get the mushrooms and you get the broccoli.

We may sometimes advise choosing secondary preferences over primary preferences. For example should I wish to quit smoking, I will have the secondary preference not to satisfy my primary preference and wish my secondary preference to prevail when the two conflict. Similarly, I may really wish to forgo my own primary preferences in order to thwart your preferences. If morality offers up normative advice how best to fulfil your non-tuistic preferences when in strategic conflict, then morality will not speak to those who eschew non-tuistic preferences for tuistic ones. Apart from stipulating the difference between preferences and metapreferences, a theory about what preferences are is beside the point. A preference-based morality needs to offer only a normative guide to resolve interpersonal conflicts over preference satisfaction. It presumes you wish to satisfy your primary preferences. It makes no assumptions about what those are save that they do not include metapreferences.[84]

3.5.4 Evolution

Once we move to an occurrent model concerning preferences and interests, we should also abandon the rational model in favour of an evolutionary model. This requires a larger argument which will be the subject of Chapters 4 and 5. A few comments here may suffice. First off, when speaking of reproductive success, we need not confine ourselves to sexual selection. Cultural beliefs may be inherited without the mediation of gene transference. If an organism has fitness due to a certain behaviour or belief (a phenotype) x, we will say that phenotype x has adaptation. That is, having x is what (or one of the features that) enables that organism to survive. So long as x is also heritable, we can predict the increase of x in that population even when *x* is not biologically inherited (a genotype). The process of cultural selection is through learning and assimilation. Learning and assimilation are the politically correct terms for the more basic drives of mimicking and habit formation. Mimicking is a built-in strategy that tends to benefit mimickers over non-mimickers.[85] Cultural inculcation is precisely the mechanism we should expect, and this returns us to the conventionality of preferences.

Notes

[1] Thomas Hobbes, *The Leviathan* (Buffalo: Prometheus Books, [1651] 1988), ch. 15, 74–5.

[2] Jeremy Bentham, *An Introduction to the Principles of Morals and Legislation* (J.H.Burns and H.L.A. Hart (eds.)) (New York: Oxford University Press, 1970), nt 4.

[3] Bernard Gert, *Morality: Its Nature and Justification* (Oxford: Oxford University Press, 1998), 112. For example, Hobbes speaks of the natural affections ("natural indulgences") of parents for their children (Thomas Hobbes, *De Cive* in Richard S. Peters (ed.) *Body, Man, and Citizen* (New York: Collier, 1962), The Second Part: "The Nature of the Body Politic", ch. 4, ¶10, 337.

[4] The objective-subjective distinction is not identical to the external-internal distinction, since we can conceive of self-interest as being the product of one's cultural group, without conceiving the cultural group's interest in objective terms. The overlap is sufficient for my purposes to equate the concepts.

[5] One such proposal is to link desire accounts to need. For a decisive critique of such an approach, see David Griffin, *Well Being: Its Meaning, Measurement and Moral Importance* (Oxford: Oxford University Press, 1986), 41–7.

[6] Alternatively, it is because she deems another interest of hers trumps the objective interest.

[7] Objective accounts of self-interest are typically associated with paternalistic doctrines, but they need not be. One might say without inconsistency, "*x* is in your interest, but it's your own affair."

[8] Joseph Butler, "Upon Human Nature," from *Fifteen Sermons Preached at the Rolls Chapel*, London, 1726. Reprinted in Bernard Baumrin (ed.) *Hobbes's Leviathan: Interpretation and Criticism* (Belmont, Calif.: Wadsworth Publishing Co., 1969), 25.

[9] Hobbes, *Leviathan*, ch. 15, 74–5.

[10] For example, Bernard Gert, *Morality: Its Nature and Justification* (Oxford: Oxford University Press, 1998), 67, 78.

[11] Hobbes, *Leviathan*, ch. 6, 24.

[12] A further worry concerns the daunting task of showing how morality serves one's subjective interests when we understand morality to be a body of social strictures on self-interest. For a discussion on this matter, see David Gauthier, *Morals by Agreement* (Oxford: Oxford University Press, 1986), ch. 6.

[13] Apart from Aristotle (*Metaphysics*, in Richard McKeon (ed.) *The Basic Works of Aristotle* (New York: Random House, 1941), XII, 6, 1072a), see also Derek Parfit, *Reasons and Persons* (Oxford: Oxford University Press, 1984), 499; Griffin, *Well Being*, 27, Susan Wolf, "Happiness and Meaning: Two Aspects of the Good Life," in E. Frankel Paul, F. Miller, Jr., and J. Paul (eds.) *Self-Interest* (Cambridge: Cambridge University Press, 1997), 211; Thomas Scanlon, *What We Owe to Each Other* (Cambridge, Mass.: Belknap, Harvard University Press, 1998), 38, 89–90; David Brink, *Moral Realism and the Foundations of Ethics* (Cambridge: Cambridge University Press, 1989), 225–6, 230; and Christine Korsgaard, *The Sources of Normativity* (Cambridge: Cambridge University Press, 1998), 50.

[14] Malcolm Murray, "Unconsidered Preferences," *South African Journal of Philosophy* 17/4 (1998): 346–53.

[15] Gert, *Morality*, 42.

[16] Parfit, *Reasons and Persons*, 496. See also his discussion of the bias toward the near, 124–5.

[17] See David Schmidtz, "Choosing Ends," *Ethics* 104 (1994): 226–51, and David Schmidtz, *Rational Choice and Human Agency* (Princeton: Princeton University Press, 1995). For an objection, see Duncan

MacIntosh, "Persons and the Satisfaction of Preferences: Problems in the Rational Kinematics of Values," *The Journal of Philosophy* 40/4 (1993): 177.

[18] Parfit's reply is also effective: "After taking certain kinds of drugs, people claim that the quality of their sensations has not altered, but they no longer dislike these sensations. We would regard such drugs as effective analgesics. This suggests that the badness of a pain consists in its being disliked, and that it is not disliked because it is bad" (Parfit, *Reasons and Persons*, 501).

[19] Simon Blackburn, *Ruling Passions* (Oxford: Oxford University Press, 2000), 80.

[20] Repetition of a negatively valenced event reduces dislike. See S.P. Grossman, *A Textbook of Physiological Psychology* (New York: John Wiley & Sons, 1973).

[21] Parfit, *Reasons and Persons*, 495.

[22] For example, Parfit, *Reasons and Persons*, 133, 464–5, Griffin, *Well Being*, 19, Gauthier, *Morals by Agreement*, 36, and Blackburn, *Ruling Passions*, 140–1.

[23] Is knowing your enemy is not destroyed really worse than falsely believing they are destroyed? Probably not *once* you learn your error. Once you learn of your mistake, you are confronted with *two* bad events: your enemy not destroyed and your having held a false belief. Two bad events are worse than one bad event, presumably. But neither bad events are *known* by you at the time. The badness is recognized at the moment you are no longer in the position of falsely believing your enemy is destroyed. The supposition that your falsely believing your enemy destroyed is worse than your knowing your enemy is not destroyed is predicated on a piece of information you cannot possibly have while you falsely believe your enemy is destroyed.

[24] See for example, Thomas Nagel's discussion about discovering one has been betrayed. It is not that the betrayal is bad because it is discovered, but its discovery makes us unhappy because betrayal is bad. Thomas Nagel, "Death" in his *Mortal Questions* (Cambridge: Cambridge University Press, 1979), 1–10.

[25] Epictetus, *Encheiridion*, in W.A. Oldfather (trans.) *Epictetus: Vol. II* (Cambridge, Mass.: Loeb Classical Library, Harvard University Press, 1996), §§ 1, 8, 14.

[26] On this point, see also Richard Brandt, *A Theory of the Good and the Right* (Amherst, N.Y.: Prometheus Books, 1998), 147–8.

[27] See, for example, Robert Nozick, *Anarchy, State, and Utopia* (New York: Basic Books, 1974), 42–5.

[28] Richard Brandt, *Facts, Values, and Morality* (Cambridge: Cambridge University Press, 1996), 42.

[29] For similar remarks, see Wayne Sumner "Something in Between," in Roger Crisp and Brad Hooker (eds.) *Well-Being and Morality: Essays in Honour of James Griffin* (Oxford University Press, 2000), 5–6.

[30] Aristotle, *Nicomachean Ethics*, in Richard McKeon (ed.) *The Basic Works of Aristotle* (New York: Random House, 1941) Bk. III, ch. 2, 1111b. See also Griffin, *Well Being*, 21–2.

[31] A danger with this definition follows from closer examination of what we can control. If our sense of control is mere conceit, then all "desires" collapse into mere "wishes" – an unhappy result. I owe this consideration to Cynthia Dennis.

[32] Parfit, *Reasons and Persons*, 498.

[33] For example, Paul Bloomfield, *Moral Reality* (Oxford: Oxford University Press, 2001).

[34] See, for example, Thomas Nagel, *The View from Nowhere* (Oxford: Oxford University Press, 1986) 143.

[35] See, for example, Thaddeus Metz, "The Reasonable and the Moral," *Social Theory and Practice*, 28 (2002): 277–301.

[36] J.S. Mill, *Utilitarianism* (Buffalo, N.Y.: Prometheus Books, [1863] 1987), 50. I reject the extension of his argument. That I desire my happiness does not commit me to desire everyone's happiness as put. I should note, perhaps unnecessarily, that while utilitarianism is often associated with hedonism, this quote does not support that. Bentham was the strict hedonist in rejecting any notion of *quality* of pleasure. Mill thought some kinds of pleasure counted more than others. A satisfied pig is not better than a disgruntled philosopher (Mill, 20).

[37] G.E. Moore, *Principia Ethica* (Cambridge: Cambridge University Press, [1903] 1986), 67.

[38] For example, Korsgaard, 253–4. See also Scanlon: "[T]o hold that [*x*] is valuable is to hold that the reasons involved in valuing it are good" (Scanlon, *What We Owe*, 90). For Raz, well-being is the

success of one's important pursuits, not just any pursuit (Joseph Raz, *The Morality of Freedom* (Oxford: Oxford University Press, 1988), 307).

[39] Brandt provides two lists of mistakes. One concerns actions, and the other concerns desires. His list of mistaken actions is: (1) overlooking options, (2) overlooking outcomes, (3) erroneous degrees of expectations, (4) errors of incremental addition, (5) discounting the future, and (6) ignoring future desires (Brandt, *The Good and the Right*, 71–87). Brandt's list of mistaken desires is: (1) dependence on false belief, (2) artificial desire-arousal in cultural transmission, (3) generalization from untypical examples, and (4) exaggerated valences produced by early deprivation (Brandt, *The Good and the Right*, 115–26). Griffin's list is: (1) factual errors, (2) logical errors, and (3) errors of insight (Griffin, *Well Being*, 11–3). Morse's list is: (1) Not accurately anticipating the effects of satisfying the desire, (2) miscalculating the costs and benefits in getting the desire, (3) mistaking how to produce the desired goods, and (4) mistaking expectations of which goods actually provide satisfaction (Jennifer Roback Morse, "Who is Rational Economic Man?" in E. Frankel Paul, F. Miller, Jr., and J. Paul (eds.) *Self-Interest* (Cambridge: Cambridge University Press, 1997), 187). Hume had only two: (1) false beliefs about facts and (2) false beliefs regarding cause and effect, both of which undermine one's choosing the appropriate means to one's desired ends, of which evidently, there could be no mistake (*Treatise*, Bk II, pt. III, sec. 3).

[40] The distinction between local and global preferences I owe to Parfit, *Reasons and Persons*, 497. See also Griffin, *Well Being*, 147. Wiggins can also be interpreted as accepting the distinction: "[W]e have to make our choice in the light of our overall practical conception of how to be or how to live a life." David Wiggins, *Needs, Values, Truth* (Oxford: Oxford University Press, 1998), 369. Even if one prefers to be spontaneous, it is conceivable to understand her actions as one driven by her global preference to be spontaneous. None of her local spontaneous actions could counteract that global preference. On this point, see Griffin, *Well Being*, 158, and Gert, *Morality*, 92.

[41] Schmidtz, "Choosing Ends." My appeal to global desires as constitutive of our personal identity is akin to Frankfurt's discussion of second-order desires (Harry Frankfurt, "Freedom of the Will and the Concept of the Person," in *The Importance of What We Care About* (Cambridge: Cambridge University Press, 1988), ch. 2), and Taylor's discussion of strong evaluations (Charles Taylor, *Human Agency and Language: Philosophical Papers 1* (Cambridge: Cambridge University Press, 1992), 16) with the qualification that, in my account, the criterion of one's strong evaluations of *worth* are not external to the agent. Perhaps closer is Rawls's account of the good being defined according to the agent's rational plan of life (John Rawls, *A Theory of Justice* (Cambridge, Mass.: Harvard University Press, 1971), 417, 433).

[42] The distinction between local and global desires is not that between instrumental and intrinsic goals, note. Local goals may be intrinsic. Desiring sexual encounters with a stranger need not be part of one's set of global goals, but may nevertheless be intrinsically satisfying. Likewise, a global goal need not be itself an intrinsic good. A global goal of studying philosophy may be instrumental to my striving for autonomy, yet still would count as a subset of my global goals. Nor are global preferences simply long-term preferences. A long-term preference may be a local preference the satisfaction of which is not for a long time off. See Raz, *The Morality*, 308.

[43] See Griffin's apt discussion, *Well Being*, 45.

[44] Parfit, *Reasons and Persons*, 498.

[45] The grass counting example stems from Rawls, *Theory of Justice*, 432–3, but see also Parfit, *Reasons and Persons*, 499–500, and Griffin, *Well Being*, 323, nt 29. For similar examples and discussion, see Brink, *Moral Realism*, 227, Wolf, *Self-Interest*, 211, Scanlon, *What We Owe*, 89–90, and Raz, *The Morality*, 301, 307–8, 316.

[46] If so, an intrinsic interest is secured by robbing the bank, no matter the outcome. Still, one may have to judge whether the thrill of the robbery outweighs the pain of the outcome.

[47] Blackburn raises a similar point in his discussion of why we do the dishes (Blackburn, *Ruling Passions*, 123). It is certainly not because we desire to do the dishes. If truth be told, we desire not to do the dishes, yet we do them anyway. Blackburn's conclusion is different from mine, though. He speaks about how obligations, predicated on what we care about, are what move us, not desires. I would say, instead, that our higher order desire in this case is cleanliness, and that is what moves us to do whatever means best accommodates that. Obligations, then, are contingent on higher type desires.

[48] William Glasser, *Reality Therapy: A New Approach to Psychiatry* (New York: Harper and Row, 1965). "Success" must be understood as a relative term in talk of inmate rehabilitation.
[49] Contra Wiggins, 239, but he is one who fails to clearly disentangle these other components.
[50] Behaviourism, or the economic view that preferences are revealed through choice, is not what I am addressing here. If behaviour reveals preference, then individual interest is revealed by choice. And if this were so, we cannot criticize choice on the grounds that it fails to maximize self-interest: it does so by definition. Although there are many reasons to dismiss behaviourism, the link between expressed attitude (itself a behaviour, by the way) and behaviour need not be discordant in the case of the grass counter or the Foole. See Gauthier, *Morals by Agreement*, 27–30.
[51] For a similar discussion, see Parfit, *Reasons and Persons*, 187, 189.
[52] For example, Sidgwick, who argued that the desirable is "what would be desired ... if it were judged attainable by voluntary action, supposing the desirer to possess a perfect forecast, emotional as well as intellectual, of the attainment or fruition" (Henry Sidgwick, *The Methods of Ethics* (Cambridge: Hackett, 7th ed., [1874] 1981), 111). See also Gauthier, *Morals by Agreement*, 23, 31; Griffin, *Well Being*, 21, 33; Parfit, *Reasons and Persons*, 118–26; Brandt, *The Good and the Right*, 46; Peter Railton, "Moral Realism," *Philosophical Review* 95/2 (1986): 163–207; and John Rawls, *Political Liberalism*, 408–24. A related approach is adopted by Scanlon. What is in your interest is not necessarily what you would endorse, but what those who care about you would endorse on your behalf (Scanlon, *What We Owe*, 4, 192). The two strands overlap when we consider that you yourself may be one of the persons you care about. This is similar to both Stephen Darwall's account ("Self-Interest and Self-Concern," in E. Frankel Paul, F. Miller, Jr., and J. Paul (eds.) *Self-Interest* (Cambridge: Cambridge University Press, 1997), 165, 176), as well as that of Rawls (*Theory of Justice*, 417).
[53] A similar complaint is made by Robert K. Shope, "The Conditional Fallacy in Contemporary Philosophy," *The Journal of Philosophy*, 75/8 (1978): 411–12. See also Darwall, 163 ("A fully knowledgeable me is not me as I am. So what a fully knowledgeable me would want is not obviously what me as I am would want"), and Griffin, *Well Being*, 11.
[54] Brandt, *The Good and the Right*, 110–26. Bloomfield, *Moral Reality*, 10.
[55] See, for example, Carl Rogers, *Client-Centered Therapy* (Boston: Houghton Mifflin, 1951).
[56] This is in line with Hobbesian contractarian analysis. See especially Gauthier, *Morals by Agreement*, 157–89. Kantian versions of contractarianism suggest the reconciliation between self-interest and morality may be achieved through an appeal to internal consistency. If I opt to be immoral while denying that option for you supposedly reveals an incoherence. See Alan Gewirth, "Can Any Final Ends Be Rational?" *Ethics* 102 (1991): 76; David Velleman, *Practical Reflection* (Princeton, Princeton University Press, 1989), 306; Kurt Baier, "Moral Reasons and Reasons to be Moral," in A.I. Goldman and J. Kim (eds.) *Values and Morals* (Dordecht: D. Reidel, 1978), 249; and Scanlon, *What We Owe*, 74. If one can happily live with such incoherence, the motivational problem is not solved. Further, that variations in normative judgements under singular and plural contexts reveals incoherence is not obvious. When I desire to win a foot race, or a lottery, I do not find it incoherent to at the same time desire your not winning. In fact, incoherence would exist if I at the same time desired both my winning and everyone else's.
[57] Richard Joyce, *The Myth of Morality* (Cambridge: Cambridge University Press, 2001), 183.
[58] Butler, *Fifteen Sermons Preached*, 20–1.
[59] Bernard Williams, "Internal and External Reasons," in Bernard Williams, *Moral Luck* (Cambridge: Cambridge University Press, 1981), 101–3. Williams's internalism holds that for *A* to have a reason to *x*, that reason must explain *A*'s doing *x*. Millgram raises a counter case. Imagine if *A* is insensitive to others and thereby has no internal reason to make friends, but friendship brings advantages that even *A* would acknowledge if only *A* were a bit more sensitive. Under such conditions, it is natural to speak of *A* having a reason to be more sensitive. Such a reason cannot, however, explain *A*'s continued insensitivity, thus Williams' internalism must be wrong. (Elijah Millgram, "Williams Argument Against External Reasons," *Noûs* (1996): 197–220.) Such a rebuttal fails, though. Insensitivity is debilitating to *A*, and *A*+ (*A* without *A*'s debilitating insensitivities) would see that *A* has a reason to become more sensitive. All this shows is that if *A* were more sensitive, *A*'s having a reason to become more sensitive would explain his action in doing so. If, on the other hand, *A* were not more sensitive, then we cannot say *A*

will be motivated to become less sensitive, and so we cannot speak of *A* having a reason to become less sensitive: we can only speak of *A+* having such a reason, but that is not *A*'s reason, however natural it may be to speak that way. For a similar reply to Millgram, see Joyce, 112–5.

[60] Jean Hampton, *The Authority of Reason* (Cambridge: Cambridge University Press, 1998), 76–8.

[61] For a more technical account of the objection, see Joyce, 115–23.

[62] "Men strive for different things. What they strive for, they call good; what they strive to avoid, they call evil." Thomas Hobbes, *The Leviathan* (Buffalo: Prometheus Books, [1651] 1988), ch. 6, 24. See also Hobbes, *De Cive*, 14.17, 166. For modern day accounts of the viability of this premise, see Owen Flanagan, *The Varieties of Moral Personality* (Cambridge, Mass.: Harvard University Press, 1991).

[63] See Christopher Morris's discussion of "fundamental evaluations" in Christopher Morris, "The Relation Between Self-Interest and Justice in Contractarian Ethics," *Social Philosophy and Policy* 5/2 (1988): 139–41.

[64] John Rawls, *Theory of Justice*, 136–42.

[65] Aside from Aristotelean doctrines (for example, Wolf, *Self-Interest*, 207–25, and David Brink, "Self-Love and Altruism," in E. Frankel Paul, F. Miller, Jr., and J. Paul (eds.) *Self-Interest* (Cambridge: Cambridge University Press, 1997), 122–57), see Gregory Kavka, "A Reconciliation Project," in D. Copp and D. Zimmerman (eds.) *Morality, Reason, and Truth* (Lanham, Md.: Rowman and Allenheld, 1984), 297–319: "[T]here are special significant pleasures or satisfactions that accompany regular moral action and the practice of a moral way of life that are not available to (unreformed) immoralists and others of their ilk."

[66] Parfit, *Reasons and Persons*, 95.

[67] David Gauthier, "Morality, Rational Choice, Semantic Representation: A Reply to My Critics," *Social Philosophy and Policy* 5/2 (1988): 192.

[68] Kurt Baier believes he has a way out of this conundrum. "Irrationality, but not contrariety to reason may depend on the manner in which preferences are held." Kurt Baier, "Rationality, Value, and Preference", *Social Philosophy and Policy*, 5/2 (1988): 42.

[69] James Fishkin, "Bargaining, Justice, and Justification: Towards Reconstruction," *Social Philosophy and Policy* 5/2 (1988): 54.

[70] Baier, "Rationality, Value, and Preference", 39. We might add to Baier's quote, "at least prior to the action." Afterward, we may deem our previous preference was not sufficiently reflected upon. But I am solely concerned with determining a preference prior to an action. I discuss post hoc deliberations below.

[71] Gauthier, *Morals by Agreement*, 34 (both quotes).

[72] Some openly accept objective tests for considered preferences. See for example, Daniel Goldstick, "Objective Interests," in H. Lyman, John Legters, P. Burke, and Arthur DiQuattro (eds.) *Critical Perspectives on Democracy* (Lanham, Md.: Roman & Littlefield, 1994), 147–64. See also Baier's argument for evaluating considered preferences by social standards (Baier, "Rationality, Value, and Preference," 44). Gauthier, on the other hand, is not in the position to admit objective standards given his commitment to a subjective account of value.

[73] See, for example, Anthony de Jasay, *Social Contract, Free Ride: A Study of the Public Goods Problem* (Oxford: Oxford University Press, 1990), 100.

[74] Gauthier, "Morality, Rational Choice, Semantic Representation," *Morals by Agreement*, 194.

[75] This information comes from my visit to the *Acadian Museum* in Miscouche, Prince Edward Island, in 2003.

[76] Joe Heinrich, Wulf Albers, Robert Boyd, Gerd Gigerenzer, Kevin McCabe, Axel Ockenfels, and Peyton Young, "What is the Role of Culture in Bounded Rationality?" in Gerd Gigerenzer and R. Selten (eds.) *Bounded Rationality: The Adaptive Toolbox* (Cambridge, Mass.: MIT Press, 2002), 353.

[77] Heinrich et al., *Bounded Rationality*, 354.

[78] Robert Boyd and Peter Richerson, *Culture and the Evolutionary Process* (Chicago: University of Chicago Press, 1985).

[79] Can't one just want to be moral? Sure, so long as this is understood as wanting to cohere with others adopting conditional cooperative behaviours, and this is not wanted for its own sake, but for the serenity, say, that social peace brings to one.

[80] Viminitz defines contractarianism as offering normative advice concerning mixed-motive interactivity. (Paul Viminitz, "What a Contractarian Must Minimally Believe about Preferences and their Revision," Canadian Association of Reductionist Philosophers Meeting, St. Mary's University, Halifax, NS, Oct. 17, 2004), 2. This is the same idea.
[81] For a critique of non-tuism, see Donald Hubin, "Non-Tuism," *Canadian Journal of Philosophy* 21/4 (1991): 441–68. For a defence of non-tuism (different than presented here), see Susan Dimock, "Defending Non-Tuism," *Canadian Journal of Philosophy* 29/2 (1999): 251–74.
[82] I would describe the Gift of the Magi problem as a looping problem, and the Battle of the Sexes game collapses into the Gift of the Magi problem when both players are moved by good tuism rather than self-interest.
[83] . . . unless your preference is to perform an immoral act, and my preference is to thwart your preference. Even still, my motive does not conform with morality, but is at least consistent with morality in that case.
[84] Robert Bass, in conversation, complains that ruling out tuistic preferences means morality cannot speak to racism and prejudice: after all, here is a case where people prefer to pursue their tuistic preferences often at the expense of their non-tuistic ones.
[85] Luca Luigi Cavalli-Sforza and Marcus Feldman, *Cultural Transmission and Evolution* (Princeton: Princeton University Press, 1981). Dan Sperba, *Explaining Culture: A Naturalistic Approach* (Oxford: Oxford University Press, 1996).

Chapter 4

RATIONALITY'S FAILURE

Hobbes's Foole asks "Why should I be moral?"[1] Gauthier's answer was that morality serves each of our interests, including the Foole's, so long as we understand our interests as our *considered* interests. The criteria for considered interests is vague, however, except that it must be consistent with Williams's internalism. I have argued (Chapter 3) that the morality-interests connection is not quite the way Gauthier described, and clarifying the relation removes the vagueness. That is, morality's role is not to serve interests, but to resolve interpersonal conflict over interests. To that extent, morality could not care less what individuals' interests are. Morality's role is purely formal, just as logic's role is purely formal in relation to premises. This qualification is not beyond what Gauthier or contract theorists can accept. The question before us now, however, is whether pointing out the conflict resolution role of morality is sufficient to convince the Foole. My answer in this chapter is to say, "No!" In the next chapter, I shall clarify why my negative answer does not lead me to moral nihilism: why my claiming that morality cannot be rationally justified does not mean I must abandon a form of contract ethics.

4.1. The Rational Model

Rational moralists argue that adhering to moral constraints is rational. Being rational is to maximize one's subjective utility. Generally it is conceded that morality's role is to thwart our rational self-interested actions, so the rational moralists must solve this apparent paradox. The motive, by the way, in identifying morality with rationality is to avoid begging the question. If we cannot justify our moral conceptions by appealing to anything other than those very moral conceptions, morality would be groundless. Those who disdain all foundationalist approaches will not have a problem with this, but moral rationalists are committed to non-question-begging reductions. In other words, rational moralists are committed to claiming that if the conditions under which it is rational to be moral are not met, so much the worse for morality.[2]

The difficulty the rational model faces is best seen as a Prisoner's Dilemma (PD), in which the rational move is defection, not cooperation. According to Hobbes, even self-interested people would be willing to do whatever it takes to escape an amoral existence. Since it will always be rational to renege on one's promises, promises to cooperate without the sword are but empty words. To solve this stalemate, Hobbes inserted a sovereign to punish defectors. Without a sovereign, the Nash Equilibrium point (the point at which any unilateral motion costs the mover) in a PD is mutual defection, but with a Sovereign who punishes defection, the Nash Equilibrium point is mutual cooperation. Under a sovereign, no one can do better than mutual cooperation by unilateral movement. Problems with the sovereign solution are many and well documented. Here are three. First, an incoherence exists. The claim is that we cannot keep agreements without the sovereign, but presumably to employ the sovereign we have to keep our agreement in electing the sovereign: an impossibility. Alternatively, we can keep our agreements without a sovereign, in which case the sovereign solution must be unnecessary. As an aside, Hobbes must accept this latter position, since he speaks of both confederacy and family bonds in the State of Nature,[3] and families may be formed and maintained "through lust."[4] A second worry with the sovereign solution entails costs. The cost of supporting a sovereign may outweigh the benefits gained. Depending on the tax rate, mutual cooperation under a sovereign may yield the same utility as mutual defection in the State of Nature. A third worry, raised by Gauthier, is that erecting a sovereign is merely a political solution, not a moral solution.[5] For a moral solution, we need to show self-constraint. We need a way of showing why internalizing the disposition to cooperate in PDs is rational. This echoes Glaucon and Adeimantus's request. Show us why the Lydian shepherd should be moral even with the ring of Gyges.

David Gauthier offers an internal solution by shifting our focus. Instead of choosing to cooperate or defect in a PD, our choice should be between adopting a disposition prone to defect and a disposition prone to cooperate only with other cooperators. The disposition prone to always defect we will call an Unconditional Defector (UD). The disposition prone to cooperate only with other cooperators we will call a Conditional Cooperator (CC).[6] To determine which disposition it is rational to adopt, we pit the two dispositions together in PD competitions and see which disposition yields the highest utility payoff for the individual players. The winner is CC. From this, Gauthier asserts that morality is rationally justified. We can ground morality without appealing to moral presuppositions.

4.2. Details

Doing better typically means getting higher scores in an interactive PD game. The moves of the game are to cooperate or to cheat. Since both you and your opponent (in a two person game) have two moves, there are four outcomes: you both cheat, you both cooperate, you cheat while the other cooperates, and you cooperate while the other cheats. Take a simple scenario: You walk into a store to buy some boots for $75.00. In such a case, you prefer boots to the money, otherwise you would not have entered the store. Meanwhile, the store owner prefers the money to the boots, otherwise he would not have opened for business. Given the potential for mutual benefit, making the exchange is deemed the contract you both make. The question becomes whether it is rational to honour the contract, or to break it. At this point, breaking the contract which you voluntarily entered seems irrational. If you think you should break it, why did you make it in the first place? Moreover, not all interactions are what de Jasay calls *spot contracts*.[7] Spot contracts are those in which the exchange happens simultaneously. The money for boots exchange is a spot contract. More often than not game situations in the real world are part-forward part-spot contracts, or "compacts," to use Hobbes's language. If Jones has already given Smith money for services to be rendered tomorrow, why should rationality not tell Smith to defect in this case? Smith is assured 4. Why settle for only 3? Things are not so simple, though.

Notice that the status quo in this case is your having the money but no boots, and the owner's keeping the boots without receiving any money. That is the state of affairs that pertained before the contract. Whatever else happens, you do not want to end up worse than your status quo. What would be worse than your status quo is to hand over the money without getting the boots in return. The owner's worst case scenario is handing over the boots without getting the money in return. Your best outcome would be to get the boots for free (the owner's worst outcome). The owner's best outcome is to get your money without handing over the boots (your worst outcome). We can arrange the outcomes in ordinal ranking according to preference satisfaction. Your ordinal ranking is as follows: first: boots, money; second: boots, no money; third: (the status quo) money, no boots; and fourth: no boots, no money. The owner's ordinal ranking is as follows: first: boots, money; second: money, no boots; third: boots, no money; and fourth: no boots, no money. In the interaction, you and the owner cannot both get your first choice, but you can both get your second choice. Rationality, however, will tell you that you can only get your third choice, and this is where the problem begins. To measure the player's utility, I shall use a reverse ordinal ranking. That is, rather than scoring 1 for your best choice, we give you 4, and rather than receiving 4 for getting your worst ranked outcome, we give you 1. Some prefer to adjust the ordinal ranking so that the status quo (the third best outcome) receives 0, but this admittedly more intuitive arrangement is not necessary, since we are only concerned with comparing players' outcomes.

The reverse ordinal ranking can be presented in the following two-by-two matrix (Table 1). In each cell, the number on the lower left represents the utility outcome for you, while the number on the upper right of each cell represents the utility outcome for the owner.

Now, so long as the owner will give you the boots on the condition that you give or will give him the money, you can get your best choice only if you deceive the owner, that is defect from the contract. Given the preference ordering of the PD, successful defection on your part yields the most points for you. The problem, one can see, is that if you were cognizant of the expected outcome values in this game, you would defect from the contract. That is, no matter what the owner does, cooperates or defects, you will do one better by defecting than by cooperating. Let us imagine that the owner cooperates. If you cooperate, you will get 3, which is good. But if you defect instead, you get 4, which is better. So, if the owner hands over the boots (cooperates), the rational thing for you to do is defect. If, on the other hand, the owner defects himself, then of your two responses (cooperating or defecting), again your most rational response is to defect. Your defecting when the owner defects gives you 2. This is better than 1, which is all you would get by cooperating while the owner defects. The result is that it is rational for you to defect no matter what the owner does.

This is an awkward conclusion for those who wish to show that it is rational to be moral. What the Prisoner's Dilemma seems to show is the reverse: immorality is what is rational. But wait! If reason tells you to

Table 1. The Prisoner's Dilemma (PD) matrix

		Owner	
		Cooperate	Defect
You	Cooperate	3 / 3	4 / 1
	Defect	1 / 4	2 / 2

defect, reason will also tell the owner to defect. What this means is that despite the mutual benefit to you and the owner that led you each to make the contract, rationality, by this picture, tells you to forget it, settle for a sub-optimal result. So the paradox of the PD is not so much that rationality cannot support morality, but rationality cannot support the best outcome. The fact that morality can support the best outcome (on the assumption that mutual cooperation is an operational definition of moral action) shows that there must be something wrong with this picture of rationality, not that there is something wrong with our picture of morality. When there is a strategy that could leave all participants better off than their respective status quos, it makes one wonder if defection is really the rational move. Recall, rationality is a tool to yield the most utiles possible, where utiles represent subjective preference satisfaction. The higher the number, the higher the degree or amount of preference satisfaction.

David Gauthier reasoned that if we take two types of players and run a number of PD games with them, the type of player that is more willing to cooperate with other cooperators is going to win out (gain more points) over one who is going to defect whenever he can. For this outcome, a limiting assumption needs to be made. As put so far, anyone who cooperates against a defector is going to do far worse than that defector. The defector will get 4 in that interaction, while the cooperator will get only 1. Although aware of this, Gauthier believed that so long as the intentions of the players were known before-hand by the other players, then it would be wise to defect against those you knew were going to defect, but cooperate with those you knew were going to cooperate. This mixed-strategy is what Gauthier refers to as a conditional strategy. One should cooperate *on the condition that* the other will cooperate; otherwise defect. This conditional move, by the way, is what makes hypothetical morality a more rationally defensible strategy than a categorical morality, which would have one cooperate no matter what the other player does.

At face value, it would seem that if you knew your opponent was going to cooperate, it would be better for you to defect, not cooperate. You would get 4 rather than the 3 given by cooperation. Gauthier circumvents this in the following way. If your disposition is to defect when the other cooperates, then other CC agents (those prone to conditionally cooperate) will see that you are prone to defect against them, and thus they will defect against you. The short-cut, as often happens in life, yields a worse result. Lured by a 4, you end up getting a 2, rather than the 3 you could have got had you been able to fend off the temptation to cheat.

Another benefit of the CC strategy is that you have no fear of ever doing worse than the status quo. So long as you can correctly identify defectors (or correctly identify defectors a certain proportion of the time), when pitted with defectors you will defect and thereby avoid your worst outcome. To adopt the CC strategy, then, means you cannot do worse, and you can only do better. As put, it seems rational to adopt the CC disposition. Given the condition of transparency (or omniscience), cheaters can expect only mutual defection, never unilateral defection, and thus will receive only 2, compared to 3 that mutual cooperation gets.

What Gauthier is highlighting is that players have not merely a choice of moves to make within the PD, but a choice of strategies. The strategy of UD is one that will defect in all cases. A UD will reason that whatever the other player does, he does better to cheat. A CC agent, on the other hand, forms the disposition to cooperate only with other cooperators. CCs will not fall prey to UDs, for CCs will detect and defect against UDs. Interactions between CCs yield rewards of 3 rather than 2 that UDs are destined to earn. The conclusion is that over repeated games with a variety of players, CCs earn more points than UDs. Thus, it is more rational to adopt the CC disposition than the UD disposition. Since Hobbes's Foole is a UD,[8] we can now show in terms of preferences that the Foole endorses (maximize your expected utility), the Foole is indeed irrational to remain a UD. People who cooperate with others who cooperate do better in the game of preference maximization than those who do not adopt this disposition. Morality is the rational choice.

For a proof of the results, let us use the following general formula for scoring utility:

$$\mu = \Sigma(u_i p_i)$$

That means that the utility of any given player using a particular strategy is the sum of the utility that player gets when playing with all other individuals using various strategies. In the case of a game where there are many players, but only the two strategies so far identified (CC and UD), we get the following more detailed scoring formula:

$$CC = 3(c - 1) + 2(d)$$

$$UD = 2[c + (d - 1)]$$

The rationale here goes as follows: The lower case letters represent the population size of each disposition in the game. The population of CCs is represented by "c," and "d" is the population of UDs. To avoid tabulating scores from a player playing against himself (a contaminant that would unfairly benefit cooperative strategies), we need to make a slight adjustment to the general formula. Hence we subtract 1 from the population of a given strategy. That is why CC gets 3(c – 1), rather than 3(c), and UD gets 2(d – 1) rather than 2(d). If a CC meets another CC, we expect 3. If the CC meets a UD, the CC will correctly identify the defector and defect, thus reaping 2. So long as there are at least two CCs, they will outperform UDs. For example, let us say there are ten UDs and two CCs in a population. Using the above formula, the scoring will be so:

$$CC = 3(2-1) + 2(10) = 26$$

$$UD = 2[2 + (10-1)] = 22$$

These results mean that in a population of ten UDs and two CCs, a CC agent can expect to earn 26 at the end of the tournament compared with any given UD's 22. Since a tournament entails playing one round with each player in the population, we can compare the average expected utility of a given CC player [2.36 (26/12 – 1)] with the average expected utility of a UD player [2 (22/12 – 1)]. A CC stands to gain 0.36 over any UD per interaction in terms of subjective utility. This shows it is rational for a UD to become a CC, or, should one be entering this population, one would do better being a CC than a UD. Morality is rational. The Foole is mistaken.

To relate this discussion back to the previous chapter, notice that the conclusion (so far) is that the Foole is mistaken by his own subjectively avowed interests. We need not try to convince him what preferences he ought to have – a good thing, since such arguments fail. Whereas, so long as his preference satisfaction hinges on outcomes of strategic interaction, he does better adopting a conditionally cooperative strategy.

By the way, CC is more rational than Rapoport's Tit-for-Tat (TFT).[9] A TFT cooperates on the first move, and afterward mimics whatever the other player did to him in their last round. A TFT meeting a UD will get 1 in the first interaction, and 2 in all future interactions with that UD. The UD will get 4 in the first interaction, and from then on 2. TFTs playing with CCs will get 3 for every interaction, as will CCs playing with TFTs. But since a CC will avoid the initial exploitation that TFT endures when playing against a UD, CCs will always outscore TFTs.

4.3. Problems

The game theoretic argument above claims that morality is rational. If it is rational, we can answer the Foole in terms the Foole should be able to endorse. Objections can come in two varieties. We can show either (1) that the CC strategy is not rational, or (2) that the CC strategy is not moral, or both. To show that CC is not rational, we can do so by showing either (i) that altering the parameters Gauthier sets for us dethrones the superiority of CC, or (ii) that CC is not a rational strategy even within the parameters that Gauthier sets for us.

1.i. The Parameters. The success of CC depends on the robustness of the conditions under which CC prevails. For CC to prevail, the following conditions must be met. (a) There are enough other CCs with whom to interact. (b) There are no extra or significant computation costs that CCs have compared to UDs. (c) The ability for CCs to correctly identify UDs is sufficiently accurate. (d) No other disposition is allowed into the game.

Other problems exist concerning game theory parameters. For example, the game theory account above assumes ordinal ranking only. The outcomes will be different if we can somehow take into account cardinal ranking of the outcomes for the players. It is true that I may value successful defection over cooperation, but by how much? And will this neatly map onto the cardinal values my competitor places on cheating and cooperating? That we can interpersonally compare ordinal rankings must never be confused with an interpersonal comparison of cardinal rankings. We know, for example that prison is a deterrent for middle-class people, but it does not appear to carry the same value for the destitute living on the streets, or for members of gangs in which prison records are a status symbol. Another problem is the lack of a significance test. Should I clearly value a score of 56 over 54 in the above scenario? The discussion here is not so much the issue between satisficing and maximizing,[10] it is an issue of the ability to distinguish significant differences. At what point should differences matter? Lastly, even with full transparency, the model is recursive, and hence unworkable. I will cooperate with you only on the condition that you will cooperate with me. But meanwhile, you will cooperate with me only on the grounds that I will cooperate with you. We seem to be at a standoff. It is more likely that we will both defect than that we will both cooperate. On a computer program, for example, the computers would not know what to do. *A*'s program says: Cooperate with *B* only if *B* cooperates with *A*. But *B* is not cooperating with *A* right now. Instead, it is waiting for *A* to cooperate with *B*. But *A* is not doing that, because it is waiting to see if *B* is going to cooperate with *A*, etc. etc.[11] Here, I shall focus only on the four problems identified in the preceding paragraph.

(a) We will get different outcomes depending on different distributions of the dispositions in the game. The formal proof indicated this is false so long as we have at least two CCs. This itself is a problem. If it is irrational to be a CC unless another CC already exists, how is the first CC to be deemed rational? This worry takes on added force once we introduce further strategies (d). The response to (a) is to note that since CCs do at least as well as UDs, the likelihood of a CC invasion increases, and this invasion will exponentially increase since the more CCs there are, the better CCs will do.

(b) CCs have computation costs that UDs do not have. After all, CCs require an extra cognitive capacity: they must inspect the other agent. UDs can be utterly blind. This added cost may be put in terms of caloric need, so that any gained utility must offset the added cost in resource use.[12] Alternatively it may be understood in terms of speed. A UD needs no time to decide what to do: it will defect. A CC, on the other hand, needs to discriminate first, and this will slow reaction time. In nature, slower reaction time may be a sufficient excuse for extinction.[13] Let us return to our population of ten UDs and two CCs, and add even a modest 10 % surcharge on CCs for computation costs per interaction. Since a 10 % tax would leave 90 % of the profit for CC, we would get the following scoring:

$$CC = (.9)(3)[(2-1)] + (.9)(2)(10) = 20.7$$

$$UD = 2[2 + (10-1)] = 22$$

CC is no longer the rational strategy given these parameters.

(c) Notice that computation costs incurred by detection apparatuses are independent of how accurate these detection apparatuses are. It is unrealistic to assume that we are never wrong in predicting how another agent will behave. Agents may vary on their ability to read others. More, agents may vary on their ability to deceive. Agents who are poor readers, or agents immersed in a pool of good deceivers, or both, will do better being a UD than a CC. Even being 90 % accurate would leave the same results noted in (b) above. Of course, sometimes CC's mistake rebounds to her own benefit. CC_1's mistaking CC_2 for a UD when CC_2 correctly identifies CC_1 as a CC (but makes the mistake of not recognizing CC_1 is making a mistake) will mean that CC_1 will get 4 in that interaction. He gains 1 in this case thanks to his proneness to make mistakes. Overall, detection errors costs CC, though. CC_2 in this case, gets only 1, and thereby loses 2 compared to what she would normally have received absent any detection errors. Since CC_1 stands to be in CC_2's place, CCs will average a one-point loss per mistake. This is the same cost incurred in the 1 % chance of two CCs simultaneously mistaking each other. It is also the same cost incurred from CC's mistaking any UDs, a far greater frequency given the greater number of UDs to CCs in the initial populations. Now, couple (b) (computation costs) and (c) (detection errors) together, and the success of CC is doubly diminished.

Of course, just as I have shown conditions in which CC fail given computation and detection costs, we can show conditions in which despite these costs, CC agents win. For example, assuming an initial population of 20 % CCs and 80 % UDs, and a 10 % error in CC's correctly identifying a UD, and another 10 % tax for computation costs, it will still be more rational to be a CC than a UD.[14] If errors in detection are set at 10 %, CCs can beat out UDs with an initial population proportion of only 10 % CCs to 90 % UDs.[15] Gauthier's answer focused on how *reputation* matters. A defector may get 4 in this encounter, but will be branded with a mark that tells future players he is a UD. Short term gain from defection will cost in the long run. Also, given the importance of correct identification to CCs, CCs will be motivated to make themselves more transparent, thus attracting like to like. Some argue that the revealing nature of body language – for example, blushing – has evolutionary success precisely for this reason.[16] We need not be transparent, so long as we are *translucent* enough to favour the CC disposition over the UD disposition.[17]

Let us grant this, and grant, too, the cases where CC wins despite heavy computation and detection costs. It is not sufficient. The reply to (a) depends on our accepting the reply to (b) and (c), and the reply to (b) and (c) depends on our accepting the reply to (a). Enough CCs can get into the mix *only if* the computation and detection costs are not penal, or so long as there are enough CCs in the mix. The problem has not gone away. How does one get enough CCs in the mix in the first place? Assuming even minimal computation and detections costs, those initial CC agents would have to be irrational: they will do worse than being a UD. CC's rationality is predicated on the irrationality of the first CC agents.

(d) What we are challenging is the rationality of CC, and the problem about new dispositions is perhaps the most serious objection. If we limit the dispositions of this game to only UD and CC, UDs can never successfully exploit others. In fact, since the UD's score is invariably 2, this is simply equivalent to a non-player. Do not forget that a score of 2 represents the status quo. By the reverse ordinal ranking, you get 2 merely for walking past the boot store. In terms of scoring, then, a UD in Gauthier's game is no different than a non-player. A UD, however, is meant to capture something different. It is neither surprising nor interesting that an agent prone to cooperate will do better than one who refuses to participate at all. What we want to know, instead, is whether an immoral agent – one who is actually able to unilaterally (i.e., successfully) defect – can do better in terms of utility than those disposed to cooperate. We need fodder for UDs. To make the UD more the agent we want to test against CC, let us introduce the exploitable into the mix. After Danielson, we will call them Unconditional Cooperators (UCs). We also need to adjust our scoring formula to accommodate the new entry. Thus:

$$CC = 3[(c-1)+u)+2(d)$$
$$UD = 4(u)+2[c+(d-1)]$$
$$UC = 3[c+(u-1)]+(1)(d)$$

That is, we assume CC cooperates with UC (for CC cooperates with anyone prone to cooperate, and UC certainly fits this bill), but defects against UD. A UD defects against everyone, of course, but since UC cooperates with everyone, UD stands to gain 4 for every interaction with UC. CC still wins this competition under many population permutations, but not all. Consider for example introducing three UCs into the population mix noted above (ten UDs, two CCs, and three UC). In this world, it pays to be a UD, not a CC.

$$CC = 3[(2-1)+3)+2(10) = 32$$
$$UD = 4(3)+2[2+(10-1)] = 34$$
$$UC = 3[2+(3-1)]+(1)(10) = 22$$

Matters get worse. Once we introduce UC, a new strategy is available, one that can do better than UD or CC. After all, if UCs are going to cooperate *no matter what*, why bother cooperating with them? Rationality, recall, is a tool to maximize expected utility given the existing parameters. But CCs fail this condition, since they could do better defecting against UCs. To show why this would be more rational, let us introduce a new disposition, whom we shall call, following Peter Danielson, a Reciprocal Cooperator (RC).[18] An RC will cooperate only on the condition that the other player's cooperation is conditional upon her own cooperation. That is, RC will cooperate with CC agents (and other RC agents), but not with UD nor – and this is the important part – with UCs. UC will cooperate no matter what. But if the reason to constrain oneself to cooperate depends on future cooperative ventures with this player, the motive to cooperate with UC vanishes. In other words, RC will exploit UCs, defect against UDs, and cooperate with CCs and other RCs. Precisely because of RCs' exploitation of UCs, RC will do better in terms of utility than CC. Now this result would not matter if we deemed RC at least as moral as CC. But since we are not likely to call moral a strategy whose success depends upon the exploitation of innocent parties, we discover a case where the rational strategy veers from the moral strategy. The desire for a rational morality fails.

For the proof, we need to adjust the previous scoring formula to take into account this new, diabolical, strategy.

$$RC = 4(u) + 3[c + (r - 1)] + 2(d)$$
$$CC = 3[(c - 1) + r + u] + 2(d)$$
$$UD = 4(u) + 2[c + r + (d - 1)]$$
$$UC = 3[c + (u - 1)] + (1)(d + r)$$

A review of the rationale for the formula may be in order. The lower case letters represent the population size of each disposition in the game. In this case, "c" is the population of CCs, "r" is the population of RCs, "u" is the population of UCs, and "d" is the population of UDs. We subtract 1 from relevant populations to avoid a player reaping points from playing with herself. If a CC meets another CC, we expect a score of 3. If the CC meets an RC, again a score of 3. Likewise with that CC's encountering a UC. But if that CC meets a UD, then she will get a score of only 2. Now if there are ten CCs, five RCs, three UDs, and two UCs in this population, the score for each disposition would be:

$$RC = 4(2) + 3(10 + (5 - 1)) + 2(3) = 56$$
$$CC = 3((10 - 1) + 5 + 2) + 2(3) = 54$$
$$UD = 4(2) + 2(10 + 5 + (3 - 1)) = 42$$
$$UC = 3(10 + (2 - 1)) + (1)(3 + 5) = 41$$

In this population mix, the most rational disposition is RC. RC does not map onto what we normally consider moral; we do not think that it is moral to rob Mother Theresa simply because she is an unconditional do-gooder. Nor do we think it moral to strike Jesus's other cheek merely because he offers it.

1.ii Rationality. Above we have challenged the rationality of CC by challenging the conditions under which CC is shown to be rational. If CC is rational only under certain conditions, and those conditions do not obtain, we have shown CC not to be rational. A more serious objection is to show that even in the best of conditions, CC is displaying an irrational strategy. As Binmore notes, if it is rational to cooperate in a PD, that shows merely that it was not a PD.[19] If the success of CC depends on initial irrationality, in what sense can we call CC agents rational? Even ignoring the problem of the first CC agent, it is always irrational to cooperate with another cooperator when unilateral defection will earn the defector greater individual utility. The rational move is precisely what CC agents prevent themselves from doing. Gauthier's response to this charge is that traditional theories of rationality have been developed within parametric choice situations: they are inapplicable under strategic choice situations. The rationality of CC, for Gauthier, is a logical extension of theories of rationality under strategic choice.[20] Peter Danielson takes another route. He argues that theories of rationality are themselves open to refutation or modification, and if CC is shown to do better than the disposition that the accepted theory of rationality would select, then that would be a reason to abandon or revise our accepted theory of rationality.[21] Both defences admit the very problem, however. Taking what we normally understand as rational, Gauthier's CC is not rational. If CC can be shown to be rational only if we alter our meaning of rationality, then this is a confession that rationality is not what we would normally ascribe to CC.

2. *Morality* Does the most rational strategy in PD games adequately represent the moral strategy? That is, even if CC were a rational strategy, and we are happy to accept the initial conditions that lead to CC's success, all is for naught if we are unlikely to accept that CC represents moral agency. This worry may be presented in two ways. First, imagine a case where Jane, who is drowning, asks passerby Dick for help. Suppose, too, that both Jane and Dick are CC agents. If Dick is a CC, we know that once Dick accepts Jane's request, Dick will be bound to carry through on his promise (assuming Dick is able to rescue Jane without harm to Dick). But merely being a CC agent does not commit Dick to agree to help Jane in the first place. Although CCs are committed to keep those agreements they make, nothing dictates they *make* them. Similarly, nothing dictates CC agents enter the boot store. They incur constraints on their actions only once they willingly enter the store. The only agreements a CC agent is willing to make are those in which she stands to benefit. If there is no benefit in entering the agreement, rationality dictates she decline. Merely being a CC does not commit one to accept the terms of every agreement proposal. Thus, Dick's being a CC does not commit Dick to accept Jane's request for help. If we want an argument that commits Dick to help Jane in this sort of case, appealing to the rational merits of CC is not equal to the task.[22] CC may capture *some* aspects of what we mean by moral agency, but not all. (Admittedly, the helping problem will not go away for evolutionary accounts either. I offer my response to this problem in Chapter 8.)

Secondly, we may return to the introduction of UC and RC. UCs are not rational since they set themselves up for exploitation by UDs. UCs will certainly do worse than CCs, but one could argue that UC better captures our understanding of moral agency than CC. After all, Jesus recommended turning the other cheek, and although we can doubt both the divinity and rationality of Jesus, few will argue that Jesus is less moral than the rest of us.[23] Matters get worse. In terms of Gauthier's and Danielson's understanding of rationality, RC agents are the most rational. For the argument that rationality and morality are integrally linked, we should expect that the most rational agent will also be the most moral. In the case of RC, this condition fails.

One might try to rescue the rational morality model by arguing, as Danielson did, that RC is more moral than either CC or UC. The existence of UCs allow UDs to prosper. Since UDs are operationally defined as immoral, UCs abet immorality. It is not uncommon to treat abetters of crimes as criminals themselves. Therefore, moral agents should punish *both* UDs and UCs.[24] Only RCs satisfy this condition. This may seem harsh. Normally punishing the innocent to get at the guilty is a miscarriage of justice. It is akin to stealing from the innocent in order to remove the temptation for thievery.[25] On the other hand, "Stop the Silence Against Violence" is a common slogan used in the campaign against violence to women. This message is aimed not merely at the abusers, but also at those who remain silent about the abuse. The moral response is to boycott all interaction with abusers, yet UCs continue to cooperate with abusive UDs. It is part of our role as moral agents to not tolerate immorality even when the immorality has no direct effect on us. This is a case where UC's "doing nothing" is immoral.

This reply strikes me as illegitimate, however. It imputes a moral motive to an action we are supposed to believe is motivated by rationality alone. Even if punishing UCs successfully rids UDs, this cannot be the motive behind RC's action. In terms of rational self-interest, RC could not possibly care whether UDs exist, since RCs are immune to UD's defection. Therefore, the motive to punish UC to get at UD is a motive beyond RC's rational arsenal.[26]

4.4. Chicken and Threat

Another argument can be made against the morality of CC. Ignoring the above problems, we might concede that CC is a moral and rational strategy in a PD. PDs are not the only sort of games people get themselves into, however. In the game of Chicken, the CC strategy (or the strategy most resembling CC) is the immoral strategy.

In the game of Chicken, two contestants race their cars toward each other. The winner is he who swerves last. I win if I keep straight while you swerve. Next best for me is if we both swerve at the same time. That would be a tie. Third best is if I swerve and you keep straight. Although I lose the game, at least I remain unharmed. The worst outcome for me is if we both keep straight and our cars ram into each other. We die or suffer terribly. Reversing the ordinal ranking, so that the higher number represents higher utility, I get 4 if I keep straight while you swerve, 3 if we both swerve at the same time, 2 if I swerve while you keep straight, and 1 if we both keep straight.

A decision matrix for the Chicken game, thus, looks like the following (Table 2):

There are two Nash Equilibrium points in this game, where you swerve and I go straight, or you go straight and I swerve. Given either of those outcomes, neither of us can do better by unilaterally altering our move. By contrast, the Nash Equilibrium point in the PD is mutual defection. Still, the mutually cooperative outcome in the Chicken game is for both to swerve, yielding 3 to each player, and this is similar to the mutually cooperative outcome of a PD, which also yields 3 to each player. One may also note the similarity between Chicken games and PDs in that in order to reap the mutually cooperative outcome, players must avoid the temptation to go straight once the other player swerves, just as a CC must avoid the temptation to defect once the other player cooperates. Appearances are deceiving, though. In a PD, a CC has the ability to threaten and perform defection should the other player defect. That is, the CC is a conditional agent. No such threat by a swerver can hold in the case of Chicken. To respond with straight rather than swerve to a straighter is clearly worse than swerving. Unlike in a PD, mutual defection (straight, straight) would yield a result worse than one's status quo. In fact, in the game of Chicken, the so-called mutually beneficial cell of both swerving is simply the status quo. Therefore, we cannot confuse the reverse ordinal ranking of 3 in Chicken with an improvement on one's status quo. The number 3 in Chicken represents one's status quo. This may lead one to suspect that the most rational strategy is not to play at all.

Table 2. The Chicken game matrix. (reverse-order ranking)

		You	
		Swerve	Straight
Me	Swerve	3 3	4 2
	Straight	2 4	1 1

Given that playing the game of Chicken is likely to yield the worst outcome for equally rational players, it can be determined in advance that playing the game is worse than not playing the game. True, you would save face if I also did not play, but you cannot be said to "lose." The slight embarrassment, at any rate, is petty compared to the expected outcome from playing, where mutual cooperation merely leaves one at one's status quo. But matters are not quite so simple. In effect, considering whether we play Chicken at all puts us in a sub-Chicken game. The decision matrix to make prior to engaging in the game of Chicken is represented in Table 3.

The idea here is that if you are even slightly embarrassed by your declining to play, would I not reap a slight victory in offering we play, on the assumption that you understand the above matrix and will thereby decline to play? I can exploit your reluctance to play a first-order Chicken game by playing a second-order Chicken game with you – even if I have neither desire nor intention to play a first-order Chicken game.

In other words, I can force you into a second-order Chicken game, and you have not the luxury of remaining at your status quo by declining. The Chicken game does not model idle games; it models the rationality behind terrorist threats. It is precisely the irrationality of your engaging in Chicken games that makes it rational for me to threaten you with Chicken games for my own profit. Because you see it is in your best interest not to play, I can use this to my advantage. Let us imagine that I have strapped explosives to my belt and I threaten to set these off in your vicinity if you do not give me $1,000. You cannot merely refuse to play in this situation. That is to be interpreted as calling my bluff. The result for failure to play, then, may be your death. Deciding whether or not to play is not an option for the threats the Chicken games are attempting to model. Threat exchanges are situations in which it costs you to fail to play. You are held at ransom. But since it costs you more to play (risk of death), I can count on your paying me the ransom for not playing. Your rationality dictates it.

Rationality tells you to acquiesce. Some may read that as proper, befitting a Ghandi or a Jesus. To seek peace is to commit oneself not to retaliate, after

Table 3. Payoffs for playing the Chicken Game

		You: Don't play	You: Play
Me	Don't play	4 4	3 2
	Play	2 3	1 1

all. For many, this is to accept the moral sacrifice. So perhaps my appeal to the game of Chicken has backfired on me – here is a game where rationality and morality coincide, after all. Alas, such a happy thought misses the depth of the Chicken game. The problem is that because I know that rationality tells you to acquiesce, and I know you are rational, rationality tells me to threaten. Rationality encourages bullies and terrorism. It pays to threaten. This is a problem for rational-based morality.

Acting on threats is not only disadvantageous to the threatened player, but also to the threatener. It is not a threat otherwise. In order to effectively threaten, the threatener must intend to do the threatened action even when doing so is not advantageous to him. This may seem a sufficient reason to call the threatener irrational. After all, how is his action rational if in order to threaten he must first commit himself to an irrational activity? This returns us to our staring point: if it is irrational to play Chicken, it is irrational to threaten to play Chicken. And knowing this, I realize that if I threaten you, it is rational for you to call my bluff. It will be rational for me at that point to chicken out. Since I know you are rational, and rationality dictates the irrationality of Chicken, I can predict you will call my bluff, so I do not want to be disposed to actually play Chicken, but since this is a requirement of effective threats, effective threats are not rational. It is not rational for me to threaten in the first place. If we cannot rationally carry out our threats, it is irrational to threaten.[27]

This argument against the rationality of threats has moved too fast, however. Insincere threats against gullible players may be quite prosperous. I may have no intention of carrying out my threat, but so long as I can convince you that I am serious, you will not likely risk calling my bluff. Moreover, if I can actually alter my disposition to categorically act on called threats, and publicize this, then my victims need no longer be merely gullible players. If self-effacement achieves my desires, rationality will in fact dictate this character transformation.[28] It is precisely this ability to self-efface that Gauthier requires in his solution to the PD. That very same ability leads one to become a committed terrorist in the game of Chicken. In other words, the CC agent is not the one to unconditionally swerve; it is she who precommits herself to go straight and to broadcast that commitment, i.e., to seek transparency.

To successfully threaten someone, you must make them believe that you will, in fact, carry out your threat. Since it is patently irrational for you to do so, you must convince the other player that you are, in fact, irrational. If I know I am encountering an irrational individual, obviously I can not trust that he will do the rational thing. If you threaten me and convince me that you will carry out your threat even if it is against your interest, then I may do better by not calling your bluff. If you can convince me, in other words, that the bottom left-hand corner of the Chicken matrix (Table 2) is an impossibility, then if I call your threat I am committing myself to accept my worst outcome, given your irrationality. In such a case, I will do better by paying you off for my not playing. Therefore it is irrational for me to call the bluff of a madman. Madmen, then, do very well in Chicken games.[29]

Can you convince me that you are irrational? Quite likely not everyone will be able to. They will have to suffer along with the rest of us without resorting to threats to get our way. The problem of threats remains, however, so long as self-effacement to a disposition conducive to irrational sincere threats is conceptually possible. Is it conceptually possible? The claim is that sometimes it is rational to be irrational.[30] If whatever furthers your preferences is rational, and being x (in this case being irrational) does this, then being x (in this case being irrational) is rational. This is peculiar, since under this guise, being irrational is rational.

Can I reasonably call your bluff? What would you do? So long as you did not somehow hard-wire yourself into acting on your threat, it would be rational for you to back down. But is it possible to hard-wire yourself in the requisite fashion? This is what Dr. Strangelove accomplished with his Doomsday machine in Stanley Kubrick's 1964 *Dr. Strangelove or How I Learned to Stop Worrying and Love the Bomb*. If it is possible, is it rational to do so knowing that you will be unable to back down when it is in your interest to do so? Of course, in the case of the Kubrick film, the error for the Peter Sellers's character was a failure to broadcast the existence of the Doomsday machine. If others are unaware he has the Doomsday machine, having the Doomsday machine *is* simply madness.

Let us say a pill is invented that, when swallowed, produces the hard-wire effect of following through with one's threats. So long as you broadcast that you have taken the pill, this should deter anyone from calling your bluff. Thus, you should have no need ever to back down yourself, since no one will ever concede to play Chicken with you. Your threat of Chicken is sufficient to get you whatever you want. After considering this, you take the pill, and will irrationally stick to your guns. You will blow us both up if I do not give you the $1,000. Assuming this is possible, might we also assume I can take a similar pill? It would not be to my advantage if I wait to transform myself only after you do so first. Two unconditional threateners competing against each other in the threat game do very badly indeed. Perhaps, however, precisely to avoid being taken advantage of in threat games, I have taken a threat-enforcer pill prior to your arrival. Now I am hard-wired to always call someone's threat; and I publicize this fact. I am the USA of foreign affairs who brashly proclaims with a poker face that I will never negotiate with terrorists. If you knew this, you would not target me. I can thereby successfully avoid threats.

Once we dream up the invention of such pills, why stop there? Let us invent yet another pill, one that disposes one to unconditionally ignore all unconditional bluff-callers. The taking of this pill must also, of course, be broadly advertised. If enough of these absolute threat masters exist in the population, it will no longer be rational to take the unconditional bluff-caller pill.

In the real world, there are psychological, biological, and sociological constraints on the kinds of dispositions we can successfully adopt. This means, unfortunately, that we may acquiesce to insincere threats. And doing this makes it rational for more insincere threateners to evolve. What is not accurately portrayed in the ordinal ranking of the matrices above, however, is that, in threat games, mutual defection (where threats are acted upon) are exceedingly costly. In many cases, the cost is far greater than the benefit from success. This needs to be factored into the formula before one can decide whether to become a sincere threatener. There will be some who decide the risk is worth it. To alter our own dispositions to counteract such defectors will quite likely be too high a price to demand. That means, unfortunately, that we must be willing to tolerate some degree of threat; and that this toleration itself creates a niche for others to capitalize on the perceived benefits of threat. It is difficult to say, at this point, where the line of toleration ends, and what precisely it is that keeps the numbers of sincere threateners low. Is it that the threshold of retaliation is itself low, or that the biological and social-psychological factors are at work to make the necessary adjustments toward rational irrationality implausible? I agree with Danielson that rational moral players must be willing to tolerate some unfairness in the world simply to avoid costly sanctions.[31]

The point of this section on threats and Chicken games is to make the case that the rational threatener in Chicken is similar to Gauthier's CC. The similarity lies in both the CC's and threatener's willingness to constrain themselves against the straightforwardly rational action. The fact that this manoeuvre may map onto moral dispositions in some games and immoral dispositions in others would seem to mean that it cannot be the rationality of conditional constraint itself that captures moral behaviour.

4.5. Summary

My purpose in this chapter was to disabuse readers of the view that morality can be rationally justified. I believe that David Gauthier's argument is the best defence of the rational morality in the history of ethics, and that is why I have focused my attention on it. Perhaps other arguments exist that readers will find better than Gauthier's, and so my conclusion that morality is not rational must be qualified to the extent that Gauthier has not shown it.

But, really, what argument *can* show such a thing? Once we admit that morality is a system of hypothetical imperatives (Chapter 2), we at the same time admit that there may be conditions under which it may be more rational, more in your interest, to not be moral. To try to argue otherwise is to deny the hypotheticity of morality.

Nevertheless, there is something right in the game theoretic approach. The problem is that we cannot look at game theory as offering normative advice about what to do in a given situation. But it can show that moral strategies are robust in a wide range of situations, and that moral strategies have evolutionary fit. But if morality has evolutionary fit, might we be tempted to say that conditional cooperative strategies (so long as that counts as moral behaviour) are justified, after all? Those who are moral do better than those who do not? As noted in Chapter 1, this is the wrong way of looking at it. Evolutionary modelling can show that it is statistically better to be moral than immoral, but this cannot provide the formal deductive proof that the Foole demands. In any event, that is what I shall argue for in the next chapter.

Notes

[1] There is a distinction between asking whether one should be moral and what is moral. Some might think the latter question needs to be decided before the former. Here, we assume things like murder and theft are immoral. Our question is why moral ascriptions about these things should concern us. And if we are unable to provide a sound reason, then so much the worse for moral ascriptions. But finding a sound answer to the "why be moral" question will in fact determine (or help determine, since empirical factors will also play a role) what counts as moral, or what besides the paradigm cases counts as moral. In this sense, the question "Why be moral?" logically precedes "What is moral?"

[2] See for example, David Gauthier, "Why Contractarianism?" in Peter Vallentyne (ed.) *Contractarianism and Rational Choice* (Cambridge: Cambridge University Press, 1991), 20.

[3] We each have strength enough to kill another through "confederacy." Thomas Hobbes,*The Leviathan* (Buffalo: Prometheus Books, 1988), ch. 13, ¶1, 63.

[4] Hobbes, *The Leviathan*, ch. 13, ¶11, 65.

[5] David Gauthier, *Morals By Agreement* (Oxford: Oxford University Press, 1988), 164.

[6] Gauthier calls the first a Straightforward Maximizer (SM) and the second a Constrained Maximizer (CM). A straightforward maximizer and an unconditional defector are not quite the same. (See Peter Danielson, *Artificial Morality: Virtuous Robots for Virtual Games*, London: Routledge, 1992, 149.) And, although a conditional cooperator is a constrained maximizer, a constrained maximizer is not necessarily a conditional cooperator (Danielson, 88–90). These differences will not matter for the ensuing discussion.

[7] Anthony de Jasay, *Social Contract, Free Ride* (Oxford: Clarendon Press, 1989), 22–3.

[8] Interestingly, the results – even if unproblematic in other respects – will not convince the Lydian shepherd, since he is not involved in any cooperative enterprise. That is a case where one can do better

on one's own. In such cases, game theoretic dynamics cannot teach us anything. Here is another reason why Glaucon and Adeimantus's challenge to Socrates has gone astray.

[9] Anatol Rapoport submitted the TFT entry in Robert Axelrod's tournament and won. See Robert Axelrod, *The Evolution of Cooperation* (New York: Basic Books, 1984), 47–8.

[10] See Schmidtz's discussion on satisficing and maximizing in David Schmidtz, "Rationality Within Reason," *The Journal of Philosophy* 89/9 (1992): 445–66.

[11] Danielson solves this problem for computer models by allowing programs to read one another's enactable algorithms. One is deemed to cooperate if one's algorithm indicates one would cooperate with a cooperator (Danielson, *Artificial Morality*, 78–84). Complications ensue if programs can put up fake algorithms (Danielson, *Artificial Morality*, 85–6).

[12] Thanks to Paul Viminitz for this picture.

[13] Thanks to Paul Thagard for this picture.

[14] Using replicator dynamics (to be discussed in the following chapter), in 74 generations, CC will take over the population: UDs will become extinct.

[15] CC will take over the population in 173 generations in this case.

[16] Richard Joyce, *The Myth of Morality* (Cambridge: Cambridge University Press, 2001), 144–5.

[17] Gauthier, *Morals By Agreement*, 174–7.

[18] Danielson, *Artificial Morality*, 12–6.

[19] "Players cannot alter the game they are playing. If it seems like they can, it is because the game has been improperly specified," Ken Binmore, *Game Theory and the Social Contract: Vol. 1: Playing Fair* (Cambridge, Mass.: MIT Press, 1998), 27. For other complaints against CC's being a "rational" strategy, see Simon Blackburn, *Ruling Passions* (Oxford: Oxford University Press, 2000), 184; Holly Smith, "Deriving Morality from Rationality," in P. Vallentyne (ed.) *Contractarianism and Rational Choice: Essays on David Gauthier's Morals By Agreement* (Cambridge: Cambridge University Press), 244–7; Brian Skyrms, *Evolution of the Social Contract* (Cambridge: Cambridge University Press, 1996), 39–40; Richmond Campbell, "Background for the Uninitiated," in R. Campbell and L. Sowden (eds.) *Paradoxes of Rationality and Co-operation* (Vancouver: University of British Columbia Press, 1985), 11. But see Danielson's response to these objections in Danielson, *Artificial Morality*, 62–3.

[20] Gauthier, *Morals By Agreement*, 183.

[21] Danielson, *Artificial Morality*, 62–3.

[22] Gauthier does not think this is a problem, by the way. First off, it just is rational to adopt a disposition to comply with agreements one thinks it is rational to make, and secondly, contractarian talk appeals to an *ex ante* agreement that encompasses all of our social rules, including the rule to be charitable. See, for example, Gauthier, "Why Contractarianism," 23. For my rebuttal, see Chapter 8.

[23] Assuming Jesus is a UC, Gauthier would say Jesus is not moral, since he believes morality requires constraint (Gauthier, *Morals By Agreement*, 2), and there is no constraint in UC agents.

[24] Danielson, *Artificial Morality*, 114–8.

[25] Gauthier's reply is similar: "If the farmer, in order to protect his chickens from the foxes, eliminates the local rabbits, thus reducing the foxes' overall food supply and encouraging them to go elsewhere, it does seem rather hard on the innocent rabbits." David Gauthier, "Moral Artifice," *Canadian Journal of Philosophy* 18 (1988): 401.

[26] On the other hand, since the existence of UC benefits UD, and the more UDs there are, the worse RC does, it might be maintained that RC's exploitation of UC is rationally motivated. I would rather say RC's selfish motive to exploit UC has a beneficial side-effect of reducing the UD population. I resurrect this argument in evolutionary terms in 5.2.2.

[27] Gauthier argues along these lines. "If it is not rational to act on one's threat because of the costs to one's life going as well as possible, it must not be rational to sincerely make the threat in the first place." David Gauthier, "Assure and Threaten," *Ethics* 104 (1994): 719. See also Jan Narveson, "The Anarchist's Case," in J. T. Sanders and J. Narveson (eds.) *For and Against the State* (Lanham, Md.: Rowman and Littlefield, 1996), 195–216.

[28] See, for example, Duncan MacIntosh, "Persons and the Satisfaction of Preferences: Problems in the Rational Kinematics of Values," *The Journal of Philosophy* 40/4 (1993): 177.

[29] For a nice account of the benefits of being mad in Chicken games, see Paul Viminitz, 'A Defence of Terrorism', in Frederick R. Adams (ed.) *Ethical Issues for the Twenty-First Century* (Charlottesville: Philosophy Documentation Center Press, 2005).
[30] Or is it rational to *appear* irrational, not *to be* irrational? Perhaps, but if the best way to appear irrational is to in fact be irrational, then the argument would lead us to become irrational.
[31] Danielson, *Artificial Morality*, 194.

Chapter 5

EVOLUTIONARY FIT

In the previous chapter, I examined the view that morality is a rational choice. Agents prone to behave in ways operationally defined as moral do better in terms of net utility maximization compared to those agents prone to behave in ways operationally defined as non-moral or immoral. I argued there that it failed. The failure of the rational model should not surprise us. Rational strategies do not generally prevail in nature. Evolution tends to favour simpler – often patently irrational – strategies. The evolutionary model of ethics maintains that *despite* their irrationality, moral agents are more likely to pass on their genes than non-moral agents. As Brian Skyrms demonstrates, evolutionary dynamics favours operationally defined moral agents who use "weakly dominated, modular irrational strategies."[1]

Admitting that CC is *not* technically rational has no impact on the evolutionary model: evolutionists never make such a claim. Evolutionary success of irrational strategies is the norm, not the exception, in nature. There are countless cases where the ideally rational move is not what evolution sells. Optimality independent of environmental and biological constraints is meaningless. In any event, the rationality of morality is not what we really need to show in order to understand the origins of morality in terms of non-moral presuppositions. The striking point of evolutionary models, as Richard Dawkins reminds us, is that it is easier to die than survive.[2] Starting with a number of possible permutations will move us to considerably fewer permutations. Since we know we do better in a moral society than in an immoral society, so long as moral agents mutate from amoral agents and manage to survive long enough to populate the world with sufficient numbers of moral agents, we can predict that the moral phenotype will carry on, and the immoral phenotype pass out of existence. The importance of turning to an evolutionary account of morality is that it could bypass the demand that the moral condition is in some sense *rational*. We can describe it as an aberration: a mutant morality. Often, evolutionary success is dependent on *not* selecting the rational choice. Luckily for us, as Hume would say, rationality is not what moves us.

5.1. Successes

Beyond Skyrms's results, the findings of Alexander, Axelrod, Binmore, and Danielson independently show how moral development has evolutionary success, despite the seeming paradox that morality is to curtail (some of) our purely self-interested actions.[3] In this section I shall summarize the general findings.

5.1.1 The Ultimatum Game

The Ultimatum Game (UG) is a game where the first player offers a division of a ten-piece pie and the second player accepts or rejects the offered division. If the second player rejects the offered division, neither player gets any of the pie.[4] The fear of getting no pie rather than some pie would make (one would think) offerers wary of seeming too greedy. Certainly an offer of "Ten for me and zero for you," stands to be rejected, leaving zero for them both.[5] But is it rational for the hungry to reject an offer of one piece of pie? If not, it would be rational to offer "Nine for me, one for you," predicting acceptance. That is, like the Prisoner's Dilemma (PD), the rational move is not the moral move. The moral move is for the first player to offer a fifty-fifty split. In this game, evolution favours the moral move.[6]

Restricting the range of offers to two: a 9-1 split and a 5-5 split, there are two possible first moves (Demand 9, Demand 5) and four possible second moves (Accept 9, Reject 9, Accept 5, Reject 5). In this game, "Accept 9" stands for accepting the demand that the other gets nine while you get one. That is, the acceptor's accepting nine really means the acceptor accepts one piece of pie. In a game where players equally swap first and second moves per interaction, we are left with eight strategies: (S1) Demand 9; Accept All. (S2) Demand 9; Reject All. (S3) Demand 9; Accept 5, Reject 9. (S4) Demand 9; Accept 9, Reject 5. (S5) Demand 5; Accept All. (S6) Demand 5; Reject All. (S7) Demand 5; Accept 5, Reject 9. (S8) Demand 5; Accept 9, Reject 5. Those who Demand 9 will do well against those who Accept 9. (Recall, "Accept 9" means accept one.) We can think in terms of Demand9ers getting nine pieces of pie rather than one, where more pie is better (no diminishing return; no cholesterol worries). How well one does depends on who else is in the mix (or who has died off, and what new agents invade). Thus "doing well" for Skyrms is translated as passing on one's genes – in this case, increasing the number of offspring for that player type. If Demand9ers do well in one generation, there will be more Demand9ers per capita in the next generation. If Demand5ers do poorly, the proportion of Demand5ers will decrease in the next generation. Each generation competes in different population mixes and these differences in population mixes will impact differently on the success of any given agent. For example, S1 (Demand 9; Accept all), will do better playing with S5 (Demand 5; Accept all) than playing with one of her own siblings (fourteen slices versus ten). S1 will do worst playing either S2 (Demand 9; reject all) or S3 (Demand 9; Reject 9, Accept 5). Notice also that for S2, S3, S6, and S8 reproductive success will backfire. These are strategies that cannot do well against their own kind. Paradoxically, the more success they have, the worse they will fare. I will return to this.

Skyrms ran 10,000 generations of this game (where every player plays every disposition per generation). Nicer players did not end up too badly so long as the initial mix favoured them. So long as we start with a 30% population majority of S7 (Demand 5; Accept 5, reject 9 (or "Fairmen" as Skyrms calls them)), immoral or unfair agents (those who Demand 9), die off. We are left with the completely obsequious S5 (Demand 5; Accept all (whom Skyrms called EasyRider)) (36%) and Fairmen (64%). The "Fairman" (S7) strategy represents moral behaviour, because it is prone to offer fair division of goods, and reject unfair divisions of goods even at a cost to the agent using that strategy. Like CC, it is willing to cooperate with others willing to cooperate, and will defect against those not willing to cooperate. "Cooperate" in this sense is understood as adhering to the convention of equal divisions of an unowned pie. If the pie were already owned by the agent, offering 50% of it to a stranger would not, one would think, be an expectation of morality. S1 or "Gamesman" captures what rationality would demand. Since it is irrational for an accepter to refuse one piece of pie (one is better than none, surely), the smart move is to offer nine. Meanwhile, if anyone else offers nine, accept it, since one piece of pie is better than none. So here is another case where (1) the rational choice is not the moral choice, and (2) where the moral strategy has evolutionary fit.[7]

5.1.2 The Narrow Bridge Game

While the Ultimatum Game may be viewed as explaining the evolution of moral equality, the Narrow Bridge Game (NBG) demands an inequality as its solution. In the NBG, two cars approach a narrow bridge from opposite ends. Only one car can cross the bridge at a time. If they both try to cross simultaneously – disaster. The solution is for one to go and one to wait. In terms of unconditional strategies, then, we will have Speeders (S) and

Waiters (W). Ss will do well against Ws, and Ws will do well against Ss, but Ss fare poorly against other Ss, and Ws fare poorly against other Ws.

The Narrow Bridge Game is the complete opposite of the Dating Game (sometimes referred to as the Battle of the Sexes Game). The Dating Game follows the vagaries of dating. A man and woman agree to go on a date together, but disagree on where to go. In stereotypic fashion, she would prefer the ballet, and he would prefer the boxing match. But they really do want to go on a date, so she would prefer to go to the boxing match with him over going to the ballet alone, and he would prefer to go to the ballet with her as opposed to going to the boxing match alone. But no amount of love can solve the dilemma, as O'Henry's *Gift of the Magi* amply shows. A handy social convention to solve the Dating Game is a rule that says: Man decides. Handy, but not fair to the woman who would always get her second choice (unless the convention has the benefit of driving her to date only men who share her passions). A fairer solution to this game is to alternate who decides, so that, in the long run, things even out. My point in raising the Dating Game, however, is to show it as a contrast to the Narrow Bridge Game. The structure of the Dating Game is that players sorely want an equal outcome, but seem doomed to an unequal result. In the Narrow Bridge Game, players desire an unequal outcome, and get frustrated by an equal result.

NBG is not the same as Chicken (discussed in 4.4). As the matrixes in the following two tables reveal, losers in the Narrow Bridge Game occur with *any* simultaneous moves. The Wait/Wait fails as poorly as the Speed/Speed (barring, perhaps, the degree of damage to the vehicles and occupants). The only palatable solution is one of inequality. Whereas in Chicken, your playing dove while the other plays dove is preferable to your playing dove while the other plays hawk, since, although you survive in both, you lose face in the latter. In the NBG, Wait/Wait amounts to a stand-off, which is worse than a Wait/Speed outcome. Playing Wait while the other Speeds permits you to carry on with your journey. There is no losing of face. Players are random passersby and there is no audience, no artificial booty. This is the difference. Perhaps talk of a new game is not needed if we can understand the game of Chicken as being embedded in a larger game with more players. Take, for example, the chariot race between Menelaus and Antilochus (Homer's *Iliad*, XXIII).[8] They race toward a narrow bridge where only one chariot can pass. It is a game of Chicken. Antilochus plays hawk and Menelaus dove. But unlike regular Chicken games, a compromise position of dove/dove is not satisfactory since Antilochus and Menelaus are not the only competitors. If they both pull up and neither go over the bridge, the other racers will pass them. That is, a dove/hawk outcome is preferable to a dove/dove outcome. This payoff captures the dynamics of the NBG.

In the individual cells of these tables (Tables 1 and 2), the number on the left of the comma represents the utility outcome for Other. The number on the right of the comma represents the utility outcome for You.

Above I said the solution of the Narrow Bridge Game demands an inequality. This is false in an important respect. The solution to any single interaction requires one to wait and the other to go: in that sense, inequality prevails. Inequality will also prevail if the only strategies allowed into this game are unconditional strategies, strategies that either simply wait (W), or simply Speed (S). But strategies exist where players can alternate roles. One strategy entails flipping a coin: heads says go, and tails says wait. Everyone's adopting a coin-flipping strategy can preserve the unequal solution in any given interaction (at least 50% of the time), while securing equality of all players, since they all adopt the same coin-flipping strategy. Since we can assume a rough equality of Ss and Ws if left to nature, just as we can assume a rough equality of Hawks and Doves in nature, shifting over to a coin-flipping strategy would not confer any advantage. Two other conditional strategies can do better than mere coin-flipping: Transposers (T) and Queuers (Q).

A Queuer (Q) follows a queuing strategy, or a "First-Come-First-Across" rule. Let us assume that each player has an equal (random) chance of meeting the criterion for being first. To follow the queuing convention is to adopt, as with Gauthier, a conditional strategy: speed if first, wait if second. Q's choice depends on the situation, not the strategy of the other player. A Transposer (T) operates like a chameleon in reverse. T waits when confronted with a Speeder, and speeds when confronted with a Waiter. T's choice depends on the strategy of the other player, not the situation.

Table 1. NBG payoffs

		You	
		Wait	Speed
Other	Wait	0,0	1,2
	Speed	2,1	−1,−1

Table 2. Chicken payoffs

		You	
		Dove	Hawk
Other	Dove	1,1	0,2
	Hawk	2,0	−1,−1

A single T will do better than a single Q in a game comprised of Ss and Ws. But Ts will not fare well against other Ts. Like Skyrms's strategies S2, S3, S6, and S8 in the Ultimatum Game, Transposers suffer from the paradox of success. No matter the difference in the initial speed of the Transposer agents, invariably they meet at the narrow bridge at the same speed. A Transposer gauges the speed of the oncoming car, and tries to adjust his or her speed accordingly. If the other car is slower, the Transposer speeds up. Alas, at the first sign of the narrows, the instinctive reaction is to slow. Thus both Transposers will gauge that the other car is slower. Thus both Transposers will simultaneously speed. Of course, if the other car is deemed to be faster, the appropriate reaction for T is now to slow. Since the other car is now perceived to increase its speed, each Transposer will now simultaneously slow. A mirrored increase and decrease of speed will continue until both Transposers collide. Transposers can do well so long as there are few Transposers. But their very doing well will violate that condition. As a result, we cannot predict an equality condition of Transposers.

Although conditional Queuers do not face the paradox of success that confronts Transposers, they face another problem. Queuers will not fare well with unconditional Speeders nor unconditional Waiters. A second-come Queuer against a Waiter will be frustrated, and a first come Queuer against a Speeder will crash. Queuers and Transposers, on the other hand, will do well when paired with one another.

I have identified four strategies in the Narrow Bridge Game: Unconditional Speeder (S), Unconditional Waiter (W), Conditional Transposer (T), and Conditional Queuer (Q). S speeds always, W slows always, T slows if the other speeds, and speeds if the other slows, and Q speeds if first at the bridge, and slows if second at the bridge. (The game is played in dyads only). Given the payoffs noted in Table 1, the following average results per interaction occurs (Table 3):[9]

The scoring average of one round favours conditional strategies (T and Q) over the unconditional strategies (W and S). As noted, both T and S

Table 3. NBG strategy utilities

	W	S	T	Q	Average
Waiter (W)	0	1	1	0.5	0.6
Speeder (S)	2	−1	2	0.5	0.9
Transposer (T)	2	1	−0.5	1.5	1.0
Queuer (Q)	1	0	1.5	1.5	1.0

suffer the paradox of success. Evolution favours Q. Assuming equal starting population proportions of individual members, replicator dynamics will drive the population to 97% Queuers, and 3% Transposers after thirty generations. Speeders and Waiters are driven to extinction.[10] So long as we conceive Unconditional Speeders as immoral, and Queuers and Transposers as two variations on moral behaviour, we have an account of why evolution favours morality. Basically, strategies which do well against each other – as moral strategies tend to do – will have better fit than strategies which fare poorly when pitted against each other – as immoral strategies tend to do.

5.1.3 Prisoner's Dilemma

So far we have shown games where the operationally defined moral strategy has more fit than operationally defined immoral strategies. As put, this may not be interesting. The criticism I raised in the last chapter was to show that the rational model fails as an answer to the PD. That the evolutionary model has success in non-PD games hardly vindicates the evolutionary model. In fact, Skyrms argues that evolution cannot support morality from a PD.[11] His thinking is that, fortunately for us, we are not in PDs (very often). If we are often in a PD, as I think we are, Skyrms's admission is a cop-out. We can do better. Skyrms is right in saying that morality cannot evolve from a PD, but only when restricting PDs to unconditional strategies. Once we introduce the conditional strategies of CC and RC from the previous chapter, evolution will come to morality's aid even in a PD.

The formula for tracking the replicator dynamics for strategy A is U(a)p(a)/**U**. U(a) = the average utility of strategy A. p(a) = the current proportion of players using strategy A. **U** = the average utility of all players in the competition (hence: $\sum[U(a_i)p(a_i)]$). Let us call $p(A)^i$ the next generation proportion for A. Then $p(A)^i = U(a)p(a)/\mathbf{U}$.

To put the formula into practice, we need a table (Table 4) charting the reverse ordinal ranking for our payoff matrix for the four strategies, UC, UD, CC, and RC, showing individual scores for row while playing against column.

Table 4. PD payoffs

	UC	UD	CC	RC	Avg
UC	3	1	3	1	2
UD	4	2	2	2	2.5
CC	3	2	3	3	2.75
RC	4	2	3	3	3

Let u, d, c, and r represent the population proportion of UC, UD, CC, and RC respectively. Using the default payoff matrix from the table above, the formula for determining the average individual scoring (U(a)) is as follows:

$$U = 3(u + c) + 1(d + r)$$
$$D = 4((u) + 2(d + c + r)$$
$$C = 3(u + c + r) + 2(d)$$
$$R = 4(u) + 3(c + r) + 2(d)$$

Notice this does not exclude self-play. We are dealing with population proportions, not raw numbers. So long as the raw numbers of players is sufficiently large, however, this omission should be negligible. Beginning with equal population proportions (0.25 for each of our four strategies), the following average utilities (U(a)) emerge:

$$U = 3(0.25 + 0.25) + 1(0.25 + 0.25) = 2$$
$$D = 4(0.25) + 2(0.25 + 0.25 + 0.25) = 2.5$$
$$C = 3(0.25 + 0.25 + 0.25) + 2(0.25) = 2.75$$
$$R = 4(0.25) + 3(0.25 + .25) + 2(0.25) = 3$$

The average utility of the population (**U**) is determined by the following formula:

$$\mathbf{U} = U(u)p(u) + U(d)p(d) + U(c)p(c) + U(r)p(r)$$

Given the equal initial population proportions, the average utility of the population (**U**) is so:

$$\mathbf{U} = 2(0.25) + 2.5(0.25) + 2.75(0.25) + 3(0.25) = 2.5625$$

Now, we have all the ingredients for the replicator dynamics to determine the population proportions for each of our dispositions in the second generation $p(A)^2$. Using our formula: $U(a)p(a)/\mathbf{U}$, gives us the following:

$$p(UC)^2 = U(u)p(u)/\mathbf{U} = 2(0.25)/2.5625 = 0.195$$
$$p(UD)^2 = U(d)p(d)/\mathbf{U} = 2.5(0.25)/2.5625 = 0.244$$
$$p(CC)^2 = U(c)p(c)/\mathbf{U} = 2.75(0.25)/2.5625 = 0.268$$
$$p(RC)^2 = U(r)p(r)/\mathbf{U} = 3(0.25)/2.5625 = 0.293$$

The cycle is repeated using these new population proportions for the subsequent generation, and the process continues until an equilibrium is reached, when $P(A)^i = P(A)$. In this case, from the assumption of equal initial population frequencies across the four dispositions (UC, UD, CC, and RC), an equilibrium is reached after seventeen generations in which CCs consume 42% of the population and RCs 58%. The results corroborate Danielson's findings noted in the previous chapter without any reliance on what counts as individually rational.

5.1.4 Summary

Two similarities among the results of PDs, UGs, and NBGs exist. One is that among the competing strategies available to the different games, conditional (or mixed) strategies do best. This returns us to the point raised in Chapter 2. Nature culls unconditional strategies. To demand that morality be unconditional, then, is to demand a strategy that nature will cull. The second similarity is that the strategy that has evolutionary fit is not the strategy that is most rational. This returns us to the betting analogy raised in my introduction. We cannot show that morality will pay for a given individual at a given time, the question the Foole demands we prove. We can only show that morality pays statistically. In each of the games examined, dispositions to coordinate with other like-minded mutants will do better than the alternatives. Morality is a good bet.

5.2. Modelling Problems and Replies

Similar problems that faced the rational model may be applied to the evolutionary model, however, and so before we rest content with the evolutionary game theory's successes, we need to show why the evolutionary model can adequately respond to these charges in ways that the rational model cannot. In Section 5.3, I shall examine problems independent of modelling constraints. In this section, I shall consider the following three problems with linking moral theory to computer simulation. (1) The problem concerning favourable initial conditions that plagued the rationality model has not been avoided by the evolutionary model. (2) The problem of whether the "winner" is really moral has also not been avoided by the shift to evolutionary models. (3) Real life is far more complex than any computer simulation can hope to simulate.

5.2.1 Favourable Conditions

The "success" of the evolutionary models occurs under very narrow conditions. In Skyrms's Ultimatum Game, for example, moral agents do better than immoral agents so long as we begin with a greater percentage of moral agents. Starting with equal population proportions will have immoral strategies win. In fact, altering Skyrms's initial population distribution by even .001 is enough to tip the scales back in favour of immoral agents. The only condition in which S7 wins is when the initial population proportions across the eight strategies are as follows: S1 = 0.32, S2 = 0.02, S3 = 0.1, S4 = 0.02, S5 = 0.1, S6 = 0.02, S7 = 0.40, S8 = 0.02. With these initial population proportions, S7 ends up with 56% of the total population along with another moral-like strategy S5 at 44%, while the rational S1 strategy goes extinct. If we alter S1 to 0.321 and S7 to 0.399, while leaving unchanged the remaining population proportions, the immoral S1 completely takes over the population in 78 generations. Skyrms has identified a condition in which fair-minded persons will have evolutionary fit, but he has also (unwittingly) shown that the odds are not in its favour. All that we can so far say, then, is that moral-*like* behaviour has evolutionary fit, or that moral behaviour has evolutionary fit in exceedingly rare cases. These are not overwhelmingly favourable results for tracking the evolution of moral behaviour.

Beyond that, the problem of self-play rears its head. Although self-play will be negligible when the population employing the given strategy is large enough, part of the problem is to account for how that population got so large in the first place. We have seen that the success of moral strategies hinges on its ability to play well with its own kind. Conversely, immoral strategies fail in the long haul because of their inability to play well with their own kind. Self-play scoring benefits only those who play well with their own kind. Therefore, failure to remove self-play benefits moral strategies over immoral strategies. If we want to show that moral strategies win despite the odds, the odds cannot be stacked in morality's favour. Failing to remove self-play stacks the odds in morality's favour.

Reply

As already noted, evolutionary models need not rely on the rationality of becoming a conditional cooperator (CC) prior to there being enough CC agents for CC to prosper. Random mutation is all the answer required.[12] Analogously, the sheer unlikeliness that *you* will win the lottery does not undermine the odds of *someone* winning the lottery. The sheer unlikeliness of earth supporting life does not undermine the odds of *some* planet supporting life. Likewise, the sheer unlikeliness of the favourable initial conditions supporting moral generation is not itself a refutation of the explanation. Why should we anticipate favourable initial mixes? "Fluke" is as sufficient an answer as we need. We do not require a reason for a mutation.

Natural selection entails the interrelation between fitness and heritability. Trait x is fitter than trait y if and only if x has a higher probability of survival and/or a greater expectation of reproductive success than y. The fitness of an organism concerns some characteristic that makes a difference in its ability to survive and reproduce.[13] Mere survival is not sufficient if the organism does not reproduce. More, the reproduction must carry the fit trait. Heritability concerns some way to ensure the offspring resemble their parents in terms of fitness. Survival in one generation must carry over to future generations before we say that trait has fitness. A problem with rational models is their too-narrow focus. It is rational to defect in a Prisoner's Dilemma. It is rational to demand nine pieces of the ten-piece pie for oneself in the Ultimatum Bargaining Game. Neither of these traits are what we would call moral strategies. But the success in terms of survival for these immoral strategies backfires in terms of heritability. The greater proportion of Defecters in the population means the fitness of Defectors decreases. Strategies that can prosper among their own kind will do better in terms of reproductive advantage than strategies that do not.

The "fluke" answer may still appear too glib for many. Another kind of reply is available. The above games rely on regulated encounters: that is, each individual player will interact with every other individual player in the competition exactly once per round. In nature, no such regulated interaction exists. Rather, interactions are, to use Aumann's language, "correlated."[14] That is, individuals interact with only a select number of other individuals. It will not matter in terms of end results whether the selection criterion is in terms of correlating with like-strategies or simple geographic convenience, or both. Such differences will affect the time it will take to reach a Nash Equilibrium. As with the pure fluke reply, though, strategies that can prosper among their own kind will do better in terms of reproductive advantage than strategies that do not. So long as interactions in nature are more correlated than random, as seems to be the case, evolutionary models of Prisoner's Dilemmas and Ultimatum Games favour moral strategies, not immoral strategies, even when the immoral strategies outnumber the moral strategies in the initial population proportions.[15] As far as the self-play problem goes, the above response suffices. So long as strategies appear through drift or mutation or recombination, and interactions are correlated, the success of moral strategies would not rely on bootstrapping scoring systems.

5.2.2 "Moral" Winners?

Evolutionary models can no more support our robust understanding of morality than the rationalist model. Although moral agents win in Skyrms's

Ultimatum Game, UCs are losers and RCs winners when applying replicator dynamics to PDs. UCs strike many as the most moral agent – the Jesus and Ghandi types who turn the other cheek when struck – and RC are deemed immoral precisely for their shameless exploitation of Jesuses and Ghandis.

Reply. Assuming equal initial population frequencies across the four dispositions (UC, UD, CC, and RC), after only seventeen generations, an equilibrium is reached in which CCs consume 42% of the population and RCs 58%. Superficially, RCs still "win," but notice that both UDs and UCs have become extinct. If UCs are extinct, our intuition against RCs' exploitation of UCs is idle. The evolutionary model predicts that we have no UCs, and so any moral indignation we feel is misdirected, or at least suspect. At any rate, without UC, RC and CC will not differ in practice. Furthermore, so long as CCs outnumber RCs and UDs outnumber UCs in the initial population mix, CCs will in fact do better than RCs. For example, if the initial population consists of 20% UCs, 30% CCs, 20% RCs, and 30% UDs, an equilibrium is reached after fifteen generations in which CCs consume 53% of the population and RCs 47%. And keeping the initial populations the same, but altering merely the differential costs of translucency between RC and CC by even .01% enables CC to come out on top of RC. (After 1000 generations, CC occupies 62% of the population, while RC occupies 38%.) Under these conditions, evolutionary fit is in fact *better* aligned with moral strategies than the rational model.

Some will say that this hardly helps the evolutionary model. In each of the games examined, including the Prisoner's Dilemma, dispositions to coordinate with other like-minded mutants will do better than the alternatives. In other words, cooperating among one's own kind has evolutionary advantage. This should not be surprising. The problem is that morality is thought to demand more than this, to cooperate even across different kinds. In fact, cooperation among one's own kind permits (or leads to) racism and sexism when irrelevant markers (such as skin colour, sex, or culturally unique body positionings) are used as indicators of cooperative strategies. So far I have given a demonstration of how rough equality can arise in specific games. If right, and generalizable, this would show that we are justified in assuming rough equality among agents in lines with the standard Hobbesian contractarian picture. But such a defence is not yet sufficient. Nothing prevents us from wondering what we ought to do should we ever come up against the easily exploitable. Saying, "Don't worry, the easily exploitable have long since been driven to extinction," does not answer *that* question. The evolutionary success of RC, for example, is not a demonstration that RC captures our full conception of morality. In other words, objectors might insist that *if* there were UCs, we *would* think it morally wrong to exploit them.[16] Therefore, the evolutionary account cannot capture morality.

We should be cautious, however, against dismissing theory whenever theory rubs against our intuitions. Our intuitions may be our problem. The moral intuition against exploiting UCs may be better understood as an overextended heuristic. Examples of overextended heuristics in nature abound. The program in a greylag goose to retrieve stray eggs to her nest can get fooled by footballs, skulls, and lightbulbs. The drive for male red-winged blackbirds to defend their territory is triggered not merely by the red wings of invading male red-winged blackbirds, but red balls and red shirts of passing joggers. The sexual attraction to bigger claws in Fiddler crabs results in offspring that are both more attractive to and less able to avoid predation. Vibrations on a web will trigger eating behaviours in female spiders that need to be circumvented by amorous male spiders (by repeated tapping on the web until the eating response is saturated). The success of a hawk in a world of doves is not carried over in a world of hawks. To say a theory of evolution must explain the greylag goose's penchant for incubating lightbulbs and the red-winged blackbird's animosity toward red balls without simply saying these are merely overextensions of otherwise useful heuristics is to demand the impossible. In cases of humans, anything involving probability reasoning is a prime example of overextended heuristics in action.[17] So long as the heuristic normally benefits agents, the heuristic will thrive, and the aberrations can be tolerated.

This is the same with some of our moral intuitions. They may well be aberrations, overextended heuristics. We may well end up with a distinction between ideal morality and a moral heuristic tracked by evolutionary models, but nothing commits us to move toward the ideal morality merely by pointing out this gap. If the "successful" disposition does not perfectly match the going account of morality, we could conclude *either* that evolutionary models have nothing to teach us about morality, *or* that our ideals of morality are overextensions of the merits of moral heuristics. Taking the evolutionary model seriously commits us to the latter. We do not say, "Because the egg retrieving heuristic has fit, there is fit in retrieving broken beer bottles." Nor should we infer that since moral dealing is good (due to its fitness) in certain circumstances, its extension to all circumstances is also good. That said, it is worth resurrecting the distinction in 4.3. RCs' exploitation of UCs does have the effect of diminishing the UD population. We cannot say that is rationally motivated, since RCs have nothing to fear from UDs, nor can they feel relative deprivation if witnessing UDs' gains from exploiting UCs. But we are no longer speaking in terms of rational motivation. All we need to observe in RCs' exploitation of UCs is the diminishing population of UDs.

The removal of UDs opens a niche for greater cooperation. Nature herself is not moral: but through the machinations of nature, a moral element arises.

5.2.3 Polymorphisms

With evolutionary models, we do not tend to arrive at a single dominant strategy. According to Skyrms, "evolutionary dynamics always carries us to a polymorphism that includes weakly dominated, modular irrational strategies."[18] None of the games examined demonstrates that polymorphism. In each case, the remaining dispositions are too similar to each other to count as a polymorphic state. In the UG, starting from a large proportion of Fairmen, we will get a polymorphic state of two moral-like dispositions (Fairmen and Easyriders). Starting from an equal proportion of all distributions gives us a polymorphic state of two immoral-like dispositions: S1 (whom Skyrms called Gamesman) (87%) and S4 (whom Skyrms called MadDog) (13%).[19] Fairmen and Easyriders do not differ in behaviour when in a population comprised of only Fairmen and Easyriders. They will both demand five and accept a demand of five. Nor will Gamesmen and MadDogs differ in behaviour when in a population consisting of only Gamesmen and MadDogs. They will both demand nine and accept a demand of nine. The end-state polymorphism in the Ultimatum Game seems artificial.[20] The same is true of the Narrow Bridge Game. True, the solution to any single interaction requires one to wait and the other to go, so in that sense, Skyrms's polymorphism prevails. But the surviving strategies are Queuers and Transposers, and we may read these as different types of moral strategies, not a moral and an immoral strategy. In the PD, the surviving dispositions are CC and RC. Absent any UCs, the differences between the two are merely formalistic. In neither the UG, the NBG, nor the PD do we get a polymorphic mix of moral and immoral surviving dispositions, but real societies do have such a polymorphic mix. Therefore, one may complain that the evolutionary model, for all its success, cannot really be representative of the real world.

I would suggest this is a constraint of modelling in general. The current state of evolutionary game theory is a closed system; the world is not. What the computer simulations show is that in closed systems, an equilibrium can be reached. But throw in a constant flux of re-entry of all possible strategies, and allow mutations of existing strategies, reaching the equilibrium is less likely. This should not defeat the conclusions reached from evolutionary modelling. The conclusion we can draw is that the trend toward a morality has evolutionary pull.

Is this an admission that morality cannot be justified to an individual, the way the rational ethics had hoped to do? The issue is like wondering whether the glass is half full or half empty. The disagreement is one more of emphasis than substance. Following from the dismissal of categoricity, we have already conceded that justification of morality in the strict sense is an impossibility. But this is not to say there is nothing to be said about morality. Here we may return to the analogy raised in my introduction. A good bet does not guarantee success. Although smoking increases the odds of getting lung cancer, some smokers may never get lung cancer. And certainly it does not indicate you will get lung cancer from your next cigarette. To prove you *will* get cancer from your next cigarette we cannot do. Our failure should not be construed as an argument for smoking, however. Similarly, the Foole demands a rational justification to be moral *now*. That we cannot do. Many variables need to be in place. There is no guarantee those variables are in place at all times for all persons. Niches for immorality exist. We may accuse the Foole of villainy, but not irrationality.[21] Our pointing out the evolutionary stability of moral dispositions is as sufficient an argument for the benefits of morality as we need.

5.3. General Problems and Replies

Apart from the specific problems with game theoretic modelling, some more general, and common, worries exist against evolutionary ethics. These worries are related, but it is helpful to separate them out. (1) Advocating irrational strategies seems patently incoherent. (2) Evolutionary ethics commit the is-ought problem. Morality is a normative notion, whereas evolutionary accounts can only speak in terms of description. (3) Related to the is-ought worry is the problem of motivation. Evolutionary models show the success of the wrong thing: heritability, not preferences. For agents who have no specific interest in the benefits of their future progeny, morality will be unmotivating. Any normative advice will fall on deaf ears. Related to this worry is the problem of free will. Evolutionary accounts may tell us how we came to have certain phenotypic traits, but so long as these phenotypic traits are not matters of choice, defectors cannot change their stripes – even if they came to see the benefits of doing so. In other words, the specific normative force of morality is lost. (4) The fourth worry concerns the shortcomings evolutionarily successful strategies have in relation to what we understand full morality to be.

5.3.1 Irrationality

Dismissing the notion of rationality may seem wildly premature. The received view of rationality entails maximizing one's personal utility. Therefore, acting rationally will secure for me more of what I prefer. So to say I might not be interested in following the advice of rationality is plainly to confess that I do not prefer that which I prefer: an incoherence.

Reply

I accept this. Still, to be rational requires brain power. To have sufficient brain power requires having the requisite brains. To have the requisite brains requires having the requisite bodies to maintain it. To have the requisite bodies, requires having the requisite resources. These are scarce, and in any event fluctuate. A cheaper model may be adequate. In nature, rationally-*like* individuals may evolve, while ideal rational agents may not. So long as certain heuristics can develop that work in normally occurring situations, the speed and convenience of cheap heuristics may benefit agents who have developed them more than agents who have not. The cost of ideal rational models may not be sufficiently compensatory, given the environments they generally inhabit. In other words, bounded rationality is all that we can hope for.[22]

A caveat is needed. A danger of appealing to simple strategies in social interactions is that they tend to be less sensitive to their external environment. Since environments are variable in both the short and the long term, plasticity is necessary if organisms are to survive to reproduce. The danger of relying too much on heuristics is to rely too much on the chances that the environment in which the heuristic works will remain. If biology teaches us anything, stable environments are not to be expected. Precisely because environments are not stable, variability is essential to biological species. This does not mean we will abandon heuristics. It is not a matter of choosing a context-insensitive heuristic over an ideal rationality. Successful heuristics will be those more sensitive to environmental fluctuations. Let us say that to be fully sensitive is to be rational in the ideal sense. But sensitivity to one's environment lies on a continuum. Fitness is a relational concept. For a given trait to have fit, it is only required to have greater fit than other available traits, not that it be supremely better. In card games, both a king and a three beats a two: to say the king beats the two *better* than the three is to misunderstand the relation. Fitness can be achieved without reaching an ideal sensitivity the way normative rational models demand.

If evolutionary fit does not favour moral development, by the way, there is nothing in the evolutionary model to justify our crying about it. As with the rational model, we would simply shrug our shoulders and say, so much the worse for morality. This may seem unbearably harsh. The difficulty in accepting that evolutionary fit should trump morality should the two conflict stems, I suspect, from the Ancients' drive to show morality has intrinsic worth. An evolutionary account can only understand moral development as having instrumental benefit, if it is to have any benefit at all. Even if a trait produces a preference such that the satisfaction of that preference yields an intrinsic benefit to the individual carrying that trait, the intrinsic benefit is itself understood in terms of its instrumental fitness. A similar conceptual snag prevents people from appreciating how damaging the problem of evil is to religion. One of the usual religious responses is that suffering, whether by sin or by natural disaster, allows us to adopt the moral response, a thing good in itself, a thing that outweighs the suffering of innocents. But the alternative is to understand morality as a successful strategy in the face of suffering, so that in the absence of suffering, we would not *require* morality. If morality is viewed as a response that benefits human agents, removing that benefit is to remove the value of morality. So although morality is not rational on the evolutionary model, morality is still seen as instrumentally beneficial – and that is as sufficient a demonstration as we need.

5.3.2 The Is-Ought Problem

Morality is a normative notion, whereas evolutionary accounts can only speak in terms of description. Evolutionary models cannot ground normative judgements at all, and so by default cannot capture morality. A descriptive account of evolutionary fitness cannot possibly provide any insight into the normativity of ethics. As even Maynard Smith observes, "A scientific theory – Darwinism or any other – has nothing to say about the value of a human being."[23] Evolution seems to show that groups are differentiated and some are destined for extinction. To infer from this that therefore it is *right* that they be so extinguished (as the Social Darwinist movement professed) is to unfoundedly assume that what *is* is also what *ought* to be. Using evolutionary models in moral discourse is a clash of concepts.

Reply

It is generally held that morality is normative: morality tells us how to behave. The naturalist must account for such normativity in natural (nonmoral) terms: a task widely thought to be impossible. Hume's Law states that no ought can be derived from an is.[24] For naturalists to have any credence, they must either reject the claim that they commit the is-ought fallacy, or deny that the is-ought fallacy is a fallacy. Searle and Anscombe, among others,[25] believe we legitimately move from ises to oughts routinely. To use Anscombe's case, as I leave the store without paying for the eggs, the store clerk will yell that the eggs cost $2.50 and that therefore I owe him

$2.50. I cannot hope to respond, "Sir, you are making an error of induction. Good day."[26] Searle notes what havoc would ensue to baseball should Hume be right about the is-ought move. A baseball player has not the option to decide whether to honour the umpire's ruling. No appeal to the is-ought problem will alter that. The description of the ball's touching the runner between bases is sufficient an account for the normative judgement that the runner is "out." The player should walk to the bench without scoring a point for the team. Does it matter that the umpire has proceeded to the normative verdict of the player's being "out" solely on the description that the player was touched by the ball between bases? Surely not, for if we cannot go from an is to an ought, the umpire's verdict is useless. Since we take the umpire's verdict as not being useless, we can go from an is to an ought.[27]

Both Searle and Anscombe are saying that despite the theoretical argument that we cannot move from an is to an ought without further ado, we do in fact move from an is to an ought without further ado. This is similar to walking over to Zeno and – in good Zen fashion – whacking him over the head when he gives his argument against motion. Hume would not be averse to such talk, having spoken in favour of nature over reason.[28] But Searle and Anscombe are mistaken. Closer inspection reveals that their cases presuppose conventional norms – or constitutive rules – and that these constitutive rules are best understood as oughts, not ises. Thus, they show we move from an is to an ought via another ought – a step that does not defeat Hume's complaint.

Take Anscombe's eggs. There is a market convention that one ought to pay the merchant his asking price if one wants his wares. Given that convention, one ought to pay the merchant his asking price if one wants his wares. And, by the way, merchants might add, you ought to so want to abide by that convention. But this added ought is not grounded in the convention itself. It is itself constitutive of the convention: that *is* the convention. This is not to say the convention is groundless. Those agents adhering to the convention may do better in terms of preference satisfaction than those agents who cannot adhere to such a convention, so long as agents who do not adhere to the convention are excluded from interacting among agents who do adhere to the convention. Is such a grounding itself a move from an is (doing better if abiding by the convention) to an ought (abide by the convention)? Only if we understand the ought as a categorical. Certainly the categorical ought cannot be grounded in an is: it cannot be grounded in anything, as we have already discussed (Chapter 2). But if we understand the ought as a hypothetical, the problem dissolves.

A hypothetical conception of morality makes the following formal claim: "If you want *x*, you should abide by rule *R*," where *R* is considered a moral rule, and *x* is something rational agents subjectively desire. So long as you want *x* and *R* gets you *x*, you ought to *R*. If "You ought to *R*" is understood categorically, agents should *R* whether or not they want *x*. Understood that way, there is a real gap in ascertaining why an agent should *R*. Understood hypothetically, the mystery is solved. The reasonableness in moving from the description to the normative claim is already comprised in our understanding what wanting *x* means. All things equal, a person wanting *x* will do what it takes to get *x*. If *R* is what it takes to get *x*, we can expect the individual to be motivated to *R*. The is-ought gap simply does not apply once we conceive morality hypothetically. Hypothetical imperatives unabashedly admit that there is no necessary connection between the is and the ought, but no one has moved from an is alone to an ought in naturalized ethics. Those with certain propensities that we call "moral dealing" tend to do better in terms of evolutionary fit than those with other propensities. That you too ought to adopt or keep that moral propensity is conditional upon your wanting to do well in these terms. Beyond hypothetical imperatives, there is no is in the world that can support any ought.

The common worry about understanding morality as (merely) a system of hypothetical imperatives is the ability to decline the invitation to *x*. We would abandon *R* if we do not want *x*, or if we do not believe abiding by *R* is the best path to *x*. If either condition is met, we have no reason to adhere to *R*. From Plato onward, this admission is thought to defeat the hypothetical conception of morality, whereas demanding anything more than this is to demand the impossible: to leap from ises to oughts. Conditionality provides the happy compromise. It is not for morality's sake that we behave morally. The worry that morality would collapse follows only if it is *also* established that adhering to morality generally does not pay. The work in evolutionary game theory says otherwise.

A cautionary note is needed. Not all features of a given species necessarily contribute to that species's survival. For example, hair growing out of one's ears in middle age does not obviously contribute to our survival. Since nature generally abhors unnecessary appendages, however, the continued existence of a strategy speaks for itself. Therefore, given the fact of our moral sentiments, the odds favour the attribution of morality playing a contributing role for our kind. Evolutionary fit is the right ordering of the right amount of a vast number of characteristics and dispositions – including social behaviours. If inherent value is something we ascribe to human beings, then it is not improbable that that ascription has evolutionary fit. To speak of the *ascription of value* to human beings is to speak about one of the features (our propensity to ascribe value to human beings) that may be part of the Homo sapiens arsenal for survival. If so, one would think evolutionary theory can account for it. After all, if the moral values we routinely ascribe to Homo sapiens have no connection to the evolutionary stability of Homo sapiens, this would need explanation.

True, evolutionary theory does not demand that everything that survives must have evolutionary fit. Fitness cannot be *defined* in terms of actual survivability while survivability is predicated on the basis of fitness. We say, instead, trait *x* is fitter than trait *y* if and only if it has a higher probability of survival and/or a greater expectation of reproductive success than *y*. But *x*'s current survival does not entail *x*'s current fitness. A trait now at fixation means only its ancestors were successful. Having fitness in one environment does not guarantee its fitness in a changed environment. And what *would* have fitness now may have become extinct before an environment favourable to it arose. Still, empirical data favours the connection between fitness and survivability with a high enough probability that the onus lies in explaining how a trait survives that has no fitness. If moral behaviour really disadvantages owners of that disposition in terms of evolutionary fit, then we should predict the weeding out of morality. Conversely, if morality has remained, we should presume our ability to track its evolutionary benefits.

Notice that when we speak of morality in this sense, we speak of cultural convention. That is where the Social Darwinists went astray. Under the Social Darwinism of Herbert Spencer,[29] for example, any non-fit allele is deemed immoral because it either fails to contribute to fitness of the group, or it is counter to fitness of the group. Thus we ought to scrap welfare programs and health programs because those who are poor or sickly are hardly fit. To have the fit take care of the unfit is to artificially bolster the continuance of unfit alleles. Whereas if the goal of evolution is to reach a pinnacle of supremely fit genes, then we should not thwart that process by artificially allowing the propagation of unfit alleles. The problem with the Social Darwinists is not (just) with making bad normative conclusions from science: they do not get their science right, either. The concept of progress in evolution must not be confused with intentionality. The norm in evolution is extinction, not a continuous direction to perfection. Rare alleles become common; common alleles become extinct.[30] There is no direction or purpose in evolution. A duty to remove that which does not fit the norm presupposes there is a norm.

The Social Darwinists made another mistake. They accord rights to those who are fit: the unfit have no rights to protect them. But this gets the picture backward. It is not the case that moral rights are accorded to those who are fit, as if morality were a prize awarded to the most deserving. Rather, those who employ conditional moral strategies are those who tend to have more fitness. In the same way, we do not accord tool use to only those with opposable thumbs. Rather, those with opposable thumbs are able to use tools, and this enhances fitness.

The Social Darwinists failed to consider our ethical beliefs as themselves being part of our evolution. To say we ought to excise the sick and infirm since they do not contribute to good offspring is to miss that we tend to find such sentiment morally repugnant. From where did this repugnance originate? It strikes me that a larger account of evolutionary dynamics should have at least *something* to say about this, and the work of Skyrms, Binmore, Alexander, and Danielson, to name a few, have tracked it in terms of conditional, strategic interaction.

5.3.3 The Motivation Problem

Evolutionary models show the success of the wrong thing: heritability, not preferences.[31] Abandoning the rational model abandons any hope of providing a justification for morality. If moral behaviour benefits me, then it is possible I will be motivated to be moral. Evolutionary accounts, on the other hand, can only speak about benefits to an organism's future offspring, not necessarily the benefits to the individual organism itself. As such, evolutionary models can only offer explanation of the development of moral behaviour in our species. And this is irrelevant to the question moral philosophers want to know: namely, is there any reason *for me* to *continue* to be moral, however beneficial morality may have been for my ancestors, or my species as a whole?

While the rational model speaks to the interests a particular agent has to be moral, the evolutionary account cannot do so. At best, evolutionary ethicists can speak to the interests of a particular agent's future generations, but this may be of little interest to the current agent. And frankly, if morality does not serve an individual agent's interests, but only its genetic reproductive success, then the same worry will apply to the agents' future offspring. That is, who I am now cannot be placated by an argument that although my interests may not be served now, my grandchildren's interests will be. Even if I somehow was concerned about my grandchildren's success over my own, the evolutionary account still would not necessarily move me, since their interests could no more be served by moral dealing under the evolutionary model than mine can. Interests in the normal intentional sense have nothing to do with reproductive success. Only a rational reduction of morality to individual preferences can hope to answer the Foole.[32]

Reply

If what I have argued in 4.4 is right, the rational model does not track preferences either. In a single PD interaction, defection will track one's preferences better than cooperation. The rationalists require one to adopt a disposition that will avoid succumbing to the temptation of always following one's occurrent preferences. Rationalists will argue that it is at least the long term benefit to the *individual* that the rationalist's model tracks, whereas it is the propagation of a *trait* that the evolution model tracks. In other words, there is a closer relation between the future individual to the present individual than grandchildren to grandparents. But as far as tracking current preferences go, the evolutionary model does better: one's preferences are part of one's heredity. The issue is no longer whether you ought to adopt a certain strategy to best track your future preferences, as if it is a matter of choice you have. The question is whether your ancestors have survived partly *because* of this hereditary trait, passed on through convention. Therefore the evolutionists can say moral behaviour actually tracks the preferences of those in which moral traits have evolved.[33] Whereas rationalists cannot say moral behaviour tracks current preferences of rational agents, only that agents would do better in the long term if they constrain preference satisfaction when under strategic choice situations – if even then.

The charge is that evolutionary modelling can provide no normative advice to the ethically challenged. My reply so far is no more than a tu quoque: I accuse the rational model of failing to provide normative advice as well. Tu quoques are not normally good moves unless the error in question is endemic or unavoidable. This is my charge. To elaborate, let me (i) clarify what the units of selection are in evolutionary modelling, and then (ii) clarify the relation between those units of selection.

i. There is some debate about what the proper unit of selection is in evolution: is it the gene, the individual organism, or the group to which the organism is a part? According to Sober and Wilson, group selectionism better explains altruism.[34] Altruism is defined as doing good for the group at risk to the individual.[35] As defined, it is difficult to explain how altruistic phenotypes could evolve if the unit of selection is the individual. When the honey bee stings the marauding bear, the barbed stinger is disembowelled from the honey bee and the honey bee is killed. Still, the sting is sufficient to scare away the bear, thus saving the hive. The barbed stinger is lethal to the individual, but does benefit the group. How could the evolution of such a deadly phenotype evolve if the individual were the unit of selection? A comparison may be made with soldiers in a fox hole. When a grenade gets tossed in, a soldier dives on the grenade, killing himself, but thereby saving the group. How could such phenotypic behaviours have evolved if the individual is the unit of selection? Sober and Wilson's answer, as well as Kropotkin's and Wynne-Edwards's,[36] is that the unit of selection is the group. Understanding that morality benefits the group is hardly the sort of answer that can satisfy the Foole. The Foole is not moved by utilitarian "proofs."[37] If evolutionary ethics gets conflated with group selectionism, my tu quoque loses force. Few biologists, however, believe that group selectionism has anything to do with evolution.

First off, the debate is not between the group and the individual, but between the group, the individual, and the gene. At the genetic level, the honey bee example poses no problem for gene selectionism, since the barbed stingers are not attached to honey bees that can have offspring. So the fact that they die off has no impact on the success of the gene, which is carried on by the reproductive success of the queen. The reproductive success of the gene in social insects is predicated on the production of sterile castes that aid non-sterile ones. The super-organism of the colony or hive is not analogous to the conglomerate of reproductive units that makes up human society. That bees, ants, and humans are each "social creatures" misses the shift in meaning of "social" across the species.

There are other examples in nature. In the case of danger calls, a bird warns the forest neighbours about a perilous intruder, and does so at risk to itself. If it simply fled the scene quietly, or even remained mute and hidden, it would do better for itself, one would think, than adopting the strategy of loudly letting the intruder know its whereabouts. In piping plovers and Wilson plovers, at least, danger calls tend to be ventriloquist acts which in fact send false or mixed signals, thereby confusing the predators. If so, the misnamed "danger call" may be explained in purely individual terms. Alternatively, the danger call stems from safety in numbers. Chickadees, for example, when they hear a danger call, swarm in groups to attack the invader. In this case, the danger call is not the selfless act; it is a call for help. The response to the danger call may be better viewed as the altruistic act. But even here, so long as there are enough collaborators, the cooperative venture is more beneficial to the individual than going out on one's own. Similarly, that mutual advantage is preferable to mutual defection in Prisoner's Dilemmas need not appeal to group selectionism. The emphasis is on *mutual* advantage, which entails each party does benefit, given the appropriate response of the other.

Showing cases where group selection is not operative does not, of course, mean group selection is never operative. True enough, but the standard examples Sober and Wilson put forward themselves fail to prove their case and this does not bode well for group selectionism. Sober and Wilson, at this juncture, point back to the strict definition of altruism and ask individual selectionists to explain the evolution of something detrimental to the individual. There are two responses. For one, genetic selection does not have the difficulty that individual selection does, and for two, one may simply say nothing satisfies the strict definition of altruism. The problem only ensues should something be found in nature to satisfy that definition. The "problem" may be with the strictness of the definition, not with gene selectionism.

Take the standard account of the evolution of morality, whether it be as I define it in terms of conditional cooperation, or as Trivers and Axelrod describes it in terms of reciprocal altruism.[38] Sober, for example, points out that reciprocal altruism is not properly altruism, since the individual reciprocator is not put at risk the way biological altruism requires.[39] *Mutual* benefit hardly counts as altruism. Altruism, standardly defined, has two conditions: (a) benefits to the other (usually in terms of reproductive fitness), and (b) expense to the self (usually in terms of reproductive fitness). Mutual benefit fails the latter condition. True enough, but can altruism strictly defined really occur in nature? Sober thinks so if we understand group selection through correlation. "What is essential is that like live with like."[40] Altruists who split off from egoists will do well on their own. The problem for Sober is that correlation is a conditional strategy. As a conditional strategy, mutual benefit is the result. Correlation would be mistaken for an unconditional strategy only so long as one treats mutual defection and non-interaction as conceptually different things. But this is implausible. Mutual defection leaves all parties at the status quo, just as non-interaction. Thus my not playing with you yields the same result as my mutually defecting with you. In this sense, correlated play is simply a kind of mixed play; a strategy that cooperates with certain kinds of individuals but not with certain other individuals. Thus cooperation is conditional upon one's co-player being "like-enough." Since correlation is itself a conditional strategy, reproductive success of correlated play is simply a case of Trivers's "reciprocal altruism," not "real" altruism, despite Sober's hopes.

A broad definition of altruism permits selection of phenotypic traits at the individual level. That is why Sober and Wilson wish to narrow the definition. Altruism narrowly defined means only the behaviours of social insects meet the criteria, and even then only when we contrast group selection with individual selection. It does not meet that condition when we understand the unit of selection as the gene. Altruism narrowly defined is a problem for Darwinian adaptationism only if we can find something that actually satisfies the definition, and this seems doubtful when one understands the unit of selection as the gene, not the individual. In any event, morality as discussed in Chapters 1, 2, and 4 concerns conditional behaviours, not altruism. Altruism narrowly defined is an unconditional act. As such, pure altruists would perish. A few mutants may pop up now and then, and we may martyrize them, but it is a confusion to picture them as offering the paradigm of moral behaviour.

The case of a soldier jumping on a grenade seems more telling. This fits the narrow definition of altruism. Here is a case where an individual clearly suffers for the good of the group. A contaminant exists in such cases, however. Soldiers are trained to put the good of the group ahead of the individual. Cultural programming can override genetic drive. This would not tell us that the unit of selection is the group any more than purposive breeding to produce a new strain of chickens shows that the unit of selection is the farmer, not the gene. When we take examples devoid of regimented training – for example, pushing a child away from an oncoming bus – the less it fits the narrow definition of altruism. The benefit to another person does not necessarily come at a cost to the benefactor. Evolutionary models need not appeal to group selectionism, contra Sober and Wilson.

This is not to say the group has no influence on selection of the gene. The group you belong to is as much a part of your environment as are available resources, competitors, and predators. In cases where individual survival is increased within groups as compared to without groups, organisms that fit into stable group dynamics will do better than those that do not, everything else being equal. To return to the African ungulates, to run slower than one's group will not bode well for you. Nor will it necessarily help to run faster than one's group. Running at the group's average running speed will be much more advantageous.

ii. If the unit of selection is neither the group nor the individual organism, but the gene, some may think this harms my case, not helps it. That moral action benefits the gene will not impress the Foole. The Foole deems himself decidedly more than the sum of his genes. The Foole demands that morality benefit the individual. But understanding the relation between the gene and the organism, or the replicator and the vehicle (or interactor),[41] dilutes this worry. A gene builds a vehicle to interact with other vehicles and the environment (or the environment which includes other vehicles, but morality's focus concerns our interaction with other vehicles, not the environment, hence my emphasis.) This means that successful vehicles entail successful genes. If the gene builds vehicles that crash, so to speak, the gene will not carry on. So, prima facie, the success of the vehicle is crucial to the success of the gene. Clearly the environment interacts directly with the vehicle, not the gene. This does not mean the vehicle is the unit of selection, for selection requires replication, and it is the gene that has the ability to do that, not the vehicle. But the direct effect on the vehicle is enough to direct our focus toward what sort of vehicles do well in what sort of environments (including vehicle populations) one is entering. So even though the individual organism's success evolutionarily speaking is not the organism itself, but the gene, the trend is that the gene requires the individual organism to do well in its environment: otherwise the gene fails to replicate.

The above is only a prima facie account. It is accurate, but only in a big picture sense. It allows a gap between the statistical and the particular, and it is in this gap where the general objections to evolutionary ethics take root. After all, the failure of a vehicle in a particular environment in a particular population at a particular time slice does not necessarily mean the failure of the gene which produces exactly that vehicle. Some phenotypes do poorly at the outset, but manage to take over the population given enough generations. Likewise, the success of some vehicles in a particular environment in a particular population at a particular time slice does not necessarily mean long term success for the gene producing that vehicle. The paradox of success (as discussed in 5.1.2) may interfere. Therefore, one may still complain that the success of the gene at the evolutionary model has no immediate or direct impact on the success of an individual organism. And since morality normally is understood to speak to individual organisms, and not one's genes, no evolutionary account of morality can ever offer normative advice to the ethically challenged. It can *describe*, but never *evaluate*, never *justify* moral behaviour in terms the Foole can understand.

Well, not so. The fact that early success need not lead to later success, nor early failure necessarily lead to later failure, is not left to mere happenstance. An underlying principle can speak to it. Concerning moral behaviour, genes that produce vehicles that (1) adopt conditional (or mixed) strategies, that (2) do well against like-strategies have greater chances of success than any genes that produce vehicles that violate either condition. These are necessary conditions, not sufficient conditions. But given this *description*, we have the sufficient means to offer sufficient *normative* advice: individuals will *tend* to do better by adhering to both principles (1) and (2). That the ultimate unit of selection is the gene, not the individual, is not germane to this point, since the success of the gene depends on the success of the phenotypic behaviour of the vehicle the gene created. We cannot be so bold as to assert a one-to-one connection, since different phenotypes may be displayed by the same gene (and different genes may display the same phenotype), but we still presume an aetiological link. Certain kinds of vehicles adopting certain kinds of phenotypic behaviours which we call "moral" tend to do better for the gene certainly, but precisely because the individual vehicle does better. Human experience occurs at the phenotypic level, not at the genetic level. But so what? So long as the vehicle displays the meme for conditional cooperation, both vehicle and gene will (tend to) prosper.

One may complain that all we can speak about is a *tendency*, and arguing for a tendency cannot provide full justification. I accept this charge, but return to my tu quoque. By the same account, no one can give a full justification. Give that up. All we can appeal to is a statistical trend, and that is not enough for Fooles if Fooles are those who demand rational justification for behaving morally in this case. Perhaps we may call them Fooles precisely for their demanding a rational justification. At any rate, we cannot call them Fooles for being irrational. But if the Foole is not speaking about a specific case, but a general strategy, then we can call them foolish, even if they periodically do better behaving immorally. Our saying it is foolish to buy a lottery ticket is not defeated by pointing out lucky winners. All we need to say is that one is foolish to count on one's luck when the odds are so much against one – statistically speaking. The same applies in the moral case. Agents adopting conditionally cooperative behaviour tend to do better in terms those agents endorse than those who do not. Letting people know this can certainly be interpreted as offering normative advice.

A related worry concerns the supposed absence of free will under evolutionary ethics. If genes are the unit of selection, then no individual can act or choose. The individual is merely a vehicle of one's genes, and genes are not altered by one's environment. To think so would be to endorse the Lamarckian picture of evolution. Rather, a gene exists, and it programmatically produces a specific vehicle and this vehicle works or does not given its environment. If it works, the gene persists. If not, the gene does not persist. So there is never any space for an individual to change stripes. By this model, we can never seriously entertain how we should morally behave: our phenotype is already encoded by our genes. This worry confuses morphological structures with behaviour. Both are phenotypic traits, but behaviours are flexible, structures are not. One is not free to choose one's jaw line, but one is free to cooperate. Nothing in the genetic selectionist account need rule out the ability of vehicles to alter behavioural strategies.

5.3.4 The Moral Shortfall

Evolutionary models indicate that morality has fit. From this we can only predict that fit under the narrow conditions in which we live. To alter any of those conditions undermines the prospect that morality may have fit. Saying this may not strike one as the right sort of move. It resurrects the complaint concerning our hypothetical treatment of UCs. It is precisely because normative talk is not apt in biology that evolutionary models are not apt in morality, say the critics. One need not deny that *is* cannot by itself imply *ought* to suggest that evolutionary models can have something to tell us about ethics. The idea that morality must have normative force may itself be a nomological dangler that should be eradicated along with the notion that moral behaviour is rational. If so, the suggestion that evolutionary models cannot accommodate what we mean by "real morality" need not bother us. After all, when we explain a conjurer's trick, the reply, "But that isn't real magic," is ignored. That we have not explained *real* magic will not bother those of us who dismiss its existence. We have explained the *semblance* of magic, and that is enough.[42] The response that the evolutionary model fails to accommodate what we *really* mean by morality should, perhaps, meet the same rebuttal.

Still, when we speak of moral norms, we speak of cultural norms. The replication of cultural norms is through custom and mimicking behaviours and various forms of social reprimand for individuals who fail to follow the norms. In this sense, then, we should be able to extrapolate a normative ethical principle from examining the common features of successful strategies in evolutionary games. Once we pay closer attention to the statistical utility of morality, we can begin to tease apart evolutionary benefits from cultural exaggerations. What people think they mean by morality and what aspect of morality is evolutionarily beneficial are not necessarily the same. As highlighted in this chapter, and consistent with our rejecting categorical moralities in Chapter 2, the bare normative advice that is consistent with the games examined would be something like: "Be a conditional cooperator!" Even for those who concede that conditional cooperation is a necessary condition of moral behaviour, given game theoretic results, many will complain that it is, in any event, not sufficient for moral behaviour. Helping others, for example, does not obviously fall out of the normative advice to be a conditional cooperator. So, something else must be at stake. This is what I shall try to offer in the following chapter.

Notes

[1] Brian Skyrms, *Evolution of the Social Contract* (Cambridge: Cambridge University Press, 1996), 32.

[2] Richard Dawkins, *The Blind Watchmaker* (Hammondsworth: Penguin, 1991).

[3] Jason Alexander, "Group Dynamics in the State of Nature," *Erkenntnis* 55 (2001): 169–82. Robert Axelrod, *The Evolution of Cooperation* (New York: Basic Books, 1984). Ken Binmore, *Game Theory and the Social Contract, Volume 1: Playing Fair* (Cambridge, Mass.: MIT Press, 1994). Peter Danielson, "Evolutionary Models of Co-operative Mechanisms: Artificial Morality and Genetic Programming," in Peter Danielson (ed.) *Modeling Rationality, Morality, and Evolution* (Oxford: Oxford University Press, 1989), 423–41.

[4] Many have wondered whether the Ultimatum Game reflects real life. In what sense would our declining an unfair portion of a public resource make the resource just disappear into thin air? Consider the following scenario. Jones has a fishing pole, but a broken arm, and so cannot fish. Smith has no broken arm, but no fishing pole, and so cannot fish. The two come upon a stream filled with trout. Smith asks Jones to lend him the pole and he will catch some fish and offer, as reward, 10% of the catch to Jones. If Jones declines, neither can get any fish.

[5] Notice how this counteracts the concept of Pareto improvements. Paretianism would demand *Accept*, since no one is worse off and at least one person is better off. But claiming that it is rational, let alone morally obligatory, to have to accept an offer of zero of a social pie merely because the offerer wants it all for himself seems insane. (If it were the offerer's pie to begin with, that would be a different matter.) For more on this, see Malcolm Murray, "Why Contractarians are not Libertarians... Evolutionarily Speaking" in Malcolm Murray (ed.) *Liberty, Games, and Contract: Jan Narveson and the Defence of Libertarianism* (Aldershot: Ashgate Press, 2007), 115–27.

[6] Does this beg the question? Why assume a 5-5 split is more *moral* than a 9-1 split? Skyrms does not recognize that someone would even *ask* this question, but libertarians certainly would. For my defence, see, "Why Contractarians are not Libertarians." For the objection, see Jan Narveson, "Social Contract, Game Theory and Liberty: Responding to My Critics," in Malcolm Murray (ed.) *Liberty, Games, and Contract: Jan Narveson and the Defence of Libertarianism* (Aldershot: Ashgate Press, 2007), 217–40.

[7] For the replicator dynamics formula that Skyrms uses, see Skyrms, 51–3. Roughly, the number of players for each strategy in the next round of games (generation) is simply the relative success of the strategy multiplied by the total number of players in the population.

[8] See Paul Dumouchel, "Rational Deception," in Caroline Gerschlager (ed.) *Deception in Markets* (Hampshire: Palgrave Macmillan, 2005), ch. 2.

[9] The scoring in the individual cells entails average payoffs for row. When T meets T, sometimes they both Speed (−1 for row), and sometimes they both slow (0 for row). Assuming the chances of Speed/Speed is equal to Wait/Wait, and the prospect of a Speed/Wait mix is nil, the average payoff for T playing a T is −0.5. In the case of Q's interaction with W, if Q arrives first, Q will Speed, and W will slow, giving Q 2 and W 1. When Q arrives second, both will slow, giving Q 0 and W 0. Assuming that each has an equal chance of arriving first, the average payoff for Q when meeting W is (2+0)/2, while the average payoff for W when meeting Q is (1+0)/2. When S meets Q, the result for S is (2 − 1)/2, while the average payoff for Q is (1 − 1)/2. When T meets Q, the average score for both is (2+1)/2. When Q meets Q, the scoring for row depends on who arrives first, and since each has an equal chance of being first, the payoff is (2+1)/2.

[10] Following Skyrms (Skyrms, 52), we can chart the replicator dynamics for strategy *A* using the formula: U(*a*)p(*a*)/**U**. U(*a*) = the average utility for a player using strategy *A*. p(*A*) = the current population proportion of players using strategy *A*. **U** = the average utility of all players. Let w, s, t, and q represent the population proportion of W, S, T, and Q respectively. The formula for individual scoring is as follows: W = 0(w)+1(s+t)+0.5(q), S = 2(w+t)+0.5(q) − 1(s), T = 2(w)+1(s)+1.5(q) − 0.5(t), and Q = 1.5(t+q)+1(w)+0(s). This does not exclude self-play. We are dealing with population proportions, not raw numbers. So long as the raw numbers of players is sufficiently large, however, this omission should be negligible. Beginning with equal population proportions (0.25 for each of our strategies), the following population proportions emerge: w = 0.179, s = 0.249, t = 0.286, and q = 0.286. Determining the population proportions for the subsequent generation proceeds as follows:

W = 0(0.179)+1(0.249+0.286)+0.5(0.286) = 0.679
S = 2(0.179+0.286)+0.5(0.286) − 1(0.249)= 0.821
T = 2(0.179)+1(0.249)+1.5(0.286) − 0.5(0.286) = 0.893
Q = 1.5(0.286+0.286)+1(0.179)+0(0.249) = 1.036
U = 0.679(0.179)+0.821(0.249)+0.893(0.286)+1.036(0.286) = 0.878
U(w)p(w)/**U** = 0.679(0.179)/0.878 = 0.14
U(s)p(s)/**U** = 0.821(0.249)/0.878 = 0.23
U(t)p(t)/**U** = 0.893(0.286)/0.878 = 0.29
U(q)p(q)/**U** = 1.036(0.286)/0.878 = 0.34

These numbers represent the population proportions for the various strategies in the next generation. After eleven generations, Waiters go extinct. After fourteen generations, Speeders go extinct. After thirty generations, Transposers represent 2.8% of the population, and the remaining 97.2% belong to Queuers. After five hundred generations, the population consists of 99% Queuers and 1% Transposers.

[11] "If prisoner's dilemmas are played in a well-mixed large population, the evolutionary dynamics drives cooperation to extinction." Brian Skyrms, *The Stag Hunt and the Evolution of Social Structure* (Cambridge: Cambridge University Press, 2004), 16.

[12] This is the extent of Alexander Rosenberg's response. "How might tit-for-tat strategies actually emerge and spread in nature, especially among organisms of limited cognitive power? The best the sociobiologist can do by way of answering this question is to point to the power of nature to provide variations in behaviour...." Alexander Rosenberg, "Altruism: Theoretical Contexts," in E. Fox Keller and E. Lloyd (eds.) *Keywords in Evolutionary Biology* (Cambridge, Mass.: Harvard University Press, 1992), 27.

[13] Elliott Sober, *Philosophy of Biology* (Boulder, Colo.: Westview Press, 1993), 70.

[14] Robert Aumann, "Subjectivity and Correlation in Randomized Strategies," *Journal of Mathematical Economics* 1(1974): 67–96. Robert Aumann, "Correlated Equilibrium as an Expression of Bayesian Rationality," *Econometrica* 55 (1981): 1–18.

[15] Skyrms, *Evolution of the Social Contract*, 63–79. Brian Skyrms and Jason Alexander, "Bargaining with Neighbors: Is Justice Contagious?" *Journal of Philosophy* 96/11(1999): 588–98. Skyrms, *Stag Hunt*, 23–5.

[16] Stephen Finlay and David Chan have independently made this objection in correspondence.

[17] See, for example, Robyn M. Dawes, *Rational Choice in an Uncertain World* (Fort Worth: Harcourt Brace College Publishers, 1988), or D. Kahneman, P. Slovic, and A. Tversky (eds.) *Judgments under Uncertainty: Heuristics and Biases* (Cambridge: Cambridge University Press, 1979).

[18] Brian Skyrms, *Evolution of the Social Contract*, 32.

[19] This matter will come up again in the problems and replies below.

[20] Similar results may be said about Axelrod's simulations. "The most successful entries tended to be relatively small variations on Tit for Tat which were designed to recognize and give up on a seemingly random player or a very uncooperative player." Robert Axelrod, *The Evolution of Cooperation* (New York: Basic Books, 1984), 47–8.

[21] See, for example Philippa Foot, "Virtues and Vices," in Philippa Foot, *Virtues and Vices* (Oxford: Oxford University Press, 2002); and also Gilbert Harman, "Moral Relativism Defended," *Philosophical Review* 84 (1975): 3–22.

[22] For an excellent discussion of bounded rationality, see Reinhard Selten, "What is Bounded Rationality?" in Gerd Gigerenzer and Reinhard Selten (eds.) *Bounded Rationality: The Adaptive Toolbox* (Cambridge, Mass.: MIT Press, 2002), 13–36. A nice overview is provided in Gerd Gigerenzer and Reinhard Selten, "Rethinking Rationality," in Gigerenzer and Selten (eds.) 1–11.

[23] John Maynard Smith, "Science and Myth," in David Hull and Michael Ruse (eds.) *The Philosophy of Biology* (Oxford: Oxford University Press, 1998), 374.

[24] In Harman's terms, we may call it the *Fact-Value* problem. In Moore's terms, we might think of it as a sub-set of the *Naturalistic Fallacy*. The naturalistic fallacy is a subset of defining one thing in terms if what it is not. It warrants the name "naturalistic fallacy" when it defines what is non-natural by something that is natural. The is-ought problem, may be construed similarly, if on the basis of some natural fact, one assumes a non-natural "ought" property can be deduced. For example, hidden in the plea: "You're hurting me, please stop!" is the presumption that the mere natural fact of being in pain is sufficient warrant for the non-natural normative judgement that one thereby ought to stop causing that pain. If normative moral judgements are not deemed non-natural, then the is-ought problem is not a naturalistic fallacy. I would read Moore as supposing this ought judgement is a simple, unanalyzable non-natural property, and that is why we might, should we so desire, speak of the is-ought problem as a subset of the naturalistic fallacy. For those of us who reject the claim that moral judgements are non-natural in this sense (as I do), then not only will we resist calling the is-ought problem a member of the naturalistic fallacy, but we will believe that the set of naturalistic fallacies is a null set.

[25] One of the others is Michael Smith, who believes the following "is" statement, "An agent gives to famine relief in circumstances C," yields the "ought" conclusion, "Giving to famine relief in circumstances C is the right thing to do," when coupled with a second premise, "Giving to famine relief in circumstances C is the feature that we would want acts to have in circumstance C if we were fully rational..." (Michael Smith, *The Moral Problem* (Oxford: Blackwell, 1994), 191) . Smith argues that the second premise is also an "is" statement, and so we have solved the is-ought problem. The second premise is not an "is" statement, though, since the phrase, "if we were fully rational" means simply what we ought to do.

[26] G.E. Anscombe, "Modern Moral Philosophy," *Philosophy* 33 (1958): 124.

[27] John Searle, "How to Derive an 'Ought' from an 'Is'," *Philosophical Review* 73 (1964): 43–58.

[28] David Hume, *A Treatise of Human Nature*, sec. II: "Of Scepticism with Regard to the Senses," L.A. Selby-Bigge (ed.) (Oxford: Oxford University Press, 1888), 187.

[29] Herbert Spencer, "Progress: Its Law and Cause," *Westminster Review* 9 (1857): 445–85. Herbert Spencer, *First Principles* (London: Williams and Norgate, 1862). For a revival of Spencer by appealing to Lamarckianism as opposed to Darwinianism, see Robert J. Richards, "The Moral Foundations of the Idea of Evolutionary Progress: Darwin, Spencer, and the Neo-Darwinians," in M.H. Nitecki (ed.) *Evolutionary Progress* (Chicago: University of Chicago Press, 1988), 129–48.

[30] "Evolution is the process by which rare alleles become common, possibly universal, and universally distributed alleles become totally eliminated." David Hull, "On Human Nature," in David Hull and Michael Ruse (eds.) *The Philosophy of Biology* (Oxford: Oxford University Press, 1998), 392.

[31] This objection has been independently raised to me in conversation by Chris Tucker, Stephen Finlay, Paul Viminitz, and an anonymous referee at Springer.

[32] See, for example, Paul Viminitz, "Getting the Baseline Right," in Malcolm Murray (ed.) *Liberty, Games and Contracts: Jan Narveson and the Defence of Libertarianism* (Aldershot: Ashgate, 2007), 129–43. Chris Tucker, in conversation, has also made similar accusations.

[33] So long as choice reveals preferences. More is needed to make this case, admittedly.

[34] See, for example, Elliot Sober and David Sloan Wilson, "A Critical review of Philosophical Work on the Units of Selection Problem," *Philosophy of Science* 61 (1994): 534–55. Elliot Sober and David Sloan Wilson, *Unto Others: The Evolution and Psychology of Unselfish Behavior* (Cambridge, Mass.: Harvard University Press, 1998).

[35] E.O. Wilson defines altruism as "self-destructive behaviour performed for the benefit of others." E.O. Wilson, *Sociobiology: The New Synthesis* (Cambridge: Cambridge University Press, 1975), 578.

[36] P. Kropotkin, *Mutual Aid: A Factor in Evolution* (London: Heinemann, 1902). V.C. Wynne-Edwards, *Animal Dispersion in Relation to Social Behaviour* (Edinburgh: Oliver and Boyd, 1962).

[37] Mill claims: "No reason can be given why the general happiness is desirable except that each person, so far as he believes it to be attainable, desires his own happiness. This, however, being a fact, we have not only all the proof which the case admits of, but all which it is possible to require, that happiness is a good: that each person's happiness is a good to that person, and the general happiness, therefore, a good to the aggregate of all persons." J.S. Mill, *Utilitarianism* (Buffalo, N.Y.: Prometheus Books, [1863] 1987), ch. 4, 50.

[38] Robert Trivers, "The Evolution of Reciprocal Cooperation," *Quarterly Review of Biology* 46 (1971): 35–57. Robert Axelrod, *The Evolution of Cooperation* (New York: Basic Books, 1984).

[39] Elliot Sober, "What is Evolutionary Altruism?" *Canadian Journal of Philosophy*, 14 (1988): 75–99. Reprinted in David Hull and Michael Ruse (eds.) *The Philosophy of Biology* (Oxford: Oxford University Press, 1998) 459–78.

[40] Sober, *Philosophy of Biology*, 473.

[41] Dawkins uses the term "vehicle" in Richard Dawkins, *The Selfish Gene* (Oxford: Oxford University Press, 1976). Hull thinks this commits one to adaptationism and since adaptationism is contentious given Gould's discussion of spandrels and exaptation (Stephen Jay Gould and Richard Lewontin, "The Spandrels of San Marco and the Panglossian Paradigm: A Critique of the Adaptationist Programme," *Proceedings of the Royal Society* B205 (1979): 581–98, and Stephen Jay Gould and Elisabeth Vrba, "Exaptation: A missing Term in the Science of Form," *Paleobiology* 8/1 (1982): 4–15), Hull suggests a more neutral term: "interactor." See David Hull, "Individuality and Selection," *Annual Review of*

Ecology and Systematics 11 (1980): 311–32. Nothing that I say here will matter or commit me to one side rather than the other, although I suspect that Gould attacks merely a very narrow and peculiar definition of adaptationism, and not one that Dawkins is committed to.

[42] Daniel Dennett offered this rebuttal to those who criticize his explanation of consciousness and free will in his Julian Jaynes Lecture: "Real Consciousness, Real Freedom, Real Magic" (University of Prince Edward Island, October, 2003). I apply it to morality.

Chapter 6

CONSENT THEORY

The betting analogy offered in the introduction is to highlight that morality cannot be rationally defended (an argument made more explicit in Chapter 4), but that it has evolutionary fit nonetheless (an argument defended in Chapter 5). But what does this tell us in terms of moral advice? In this chapter, I argue that a similarity exists between the successful strategies in various games, and this will have something to do with conditional cooperation. From this, we can extrapolate the following normative rule: *Do not do unto others without their consent.* I shall call this the *principle of consent.* At first view, the principle of consent will not appear to be a conditional strategy, but any other player, *B*, who rejects the principle of consent releases any player, *A*, from any wrongdoing toward *B*. By rejecting the principle of consent, they consent to any act done them. In turn, this shows the merit of accepting the principle of consent. But, given the complexity of strategic interaction, such an argument falls short of providing rational justification for *B* to accept the principle of consent. The normative rule is neither objectively true, nor rationally justified, nor broad in scope.

We understand the claim that a woman gives birth every seven seconds as a statistical claim, not a claim about a particular woman. It is unjustifiable as a claim about any given woman. Noting this does not dampen our belief that the claim is justifiable (or may be justifiable) in the statistical sense. Do we say, therefore, that it is unjustifiable? Hardly. To say it is an unjustifiable utterance is to say we *intend* it as a claim about a particular woman. That is where the error occurs. When we intend it as a statistical generalization, the claim is justifiable to the extent that it fairly represents the average.

Of course, how often women give birth is a descriptive claim, so some further remark needs be made about the justifiability about normative claims, like "Don't steal!" But this is done to the extent that, given their social circumstances, people statistically do better if they do not steal than if they do. And if this is driven by one's genes (please note the "if"), those genes which produce vehicles that adopt conditionally cooperative phenotypic strategies, then, everything else being equal, those genes will prosper. The insertion of "in general" is to tolerate cases to the contrary. Pointing out a case of a woman never giving birth, or a woman giving birth only once within a year hardly counters the statistical generalization that the number of human births per year is equal to one every seven seconds. That people typically get lung cancer from smoking is not refuted by pointing out a few cases of smokers who never get lung cancer. Similarly, pointing out a case where one can do better stealing than not stealing does not upset the normative claim, taken as a statistical generalization. This is why Glaucon and Adeimantus's demand of Socrates is ill-grounded. We cannot hope to show that morality is worthwhile in every single case. Perhaps it cannot be shown to be worthwhile in any single case. None of this affects the statistical, or evolutionary, benefits of morality. The addition of a single grain of sand can never make a heap. The odds of getting cancer from a single cigarette are low. Neither fact settles the matter. It shows merely the myopic application of rationality. Evolution, like statistical analysis, looks at a broader time span. Evolution, not rationality, moves us. This is all the "proof" required. How does evolution move us? It provides us with a normative heuristic sufficiently captured by the principle of consent.

The principle of consent is a variant of contractarianism. Reliance on evolutionary support mandates certain structural changes to the contractarian picture, however. My plan in this chapter is to clarify and to support these alterations, not to abandon contractarianism. In fact, much of what I say will concur with what Gauthier and Kavka say. For example, Kavka points out how it is a poor gamble to violate moral rules,[1] and Gauthier appeals to the overall – not the short-term – benefits by adopting a disposition to conditionally cooperate.[2] With these claims I agree. My complaint is that recognizing these truths is one thing, rationally committing oneself to be bound by moral constraints another. Kavka, perhaps more uneasy than Gauthier with the rational solution to the compliance gap, argues that rationality is all that is required for "those already endowed with conscience and moral motivations."[3] Convenient. On the evolutionary model, we avoid the compliance problem. We get a natural reductionist, contractarian picture without claiming moral motives are rationally derived.

6.1. The Simplified Normative Principle of Consent

Results from game theory show that evolutionarily successful strategies are conditional ones, not unconditional ones. This helps support what we have already inferred after a conceptual analysis in Chapter 2, that morality is hypothetical not categorical. As shown in Chapter 5, those disposed to cooperate with other cooperators, those disposed to offer fair divisions of goods and to reject unfair divisions of goods, and those willing to queue rather than barge ahead of others, have evolutionary fit. In terms of normative advice, then, we might tell those we care about, including ourselves, that we should do likewise. To put that into a more general principle is to present a form of contract theory, or so I shall attempt to argue here. I shall offer a naturalized version of contractarianism. In brief, game theory helps to highlight the evolution of morality as a resolution of interpersonal conflicts under strategic negotiation. It is this emphasis on strategic negotiation that underwrites the idea of consent.

A simplified version of the consent principle can be put in the negative: *Don't do to others without their consent.* This does not quite cover all cases. The full version would need to add, "*. . . or their expected consent, or what they would consent to if they had sufficient cognitive capacity.*" Demanding that we should at least *expect* their consent if we are to do anything to them covers falling piano and surprise party cases. My pushing Sally out of the way is not what she would normally consent to, but so long as we can reasonably expect she would consent to my pushiness once she discovers that was the best means of saving her from a falling piano, then my action meets consent theory's approval. We cannot generalize from this case, however. Determining expected consent is not a piece of cake. People differ, and what one might consent to another might not. If we know that Jean loves surprise parties and Gertrude hates them, then throwing a surprise party for Jean is permissible while throwing a surprise party for Gertrude is not. When one wrongly assumes expected consent, we can wonder whether it was reasonable to have expected consent. If so, the condemnation is weaker. The criteria for determining expected consent is fuzzy. It gets fuzzier once we move to making consent decisions for the incapacitated. In the surprise party case, we can benefit from a post-experiment follow-up. After we throw the surprise party and discover Gertrude's annoyance with it, we learn not to throw another surprise party for Gertrude, and perhaps also learn to better recognize clues indicating who does not like surprise parties, so that we will not make that mistake with others. Learning in this way requires follow-up communication, but follow-up communication may be as impossible as pre-communication for the long-term or permanently incapacitated. Still, the idea is to permit doing *good* things to those who cannot, and perhaps may never, speak for themselves. If they were cognitively capable and would consent, then presumably it is not a bad thing we are doing to them. The difficulty, of course, is in determining whether someone *would* consent. At this level, we appeal to what reasonable people tend to consent to in similar circumstances. The fuzziness has not been eluded, but my concern lies elsewhere.

Many contract theorists believe that since we are allowed to move to this hypothetical domain in judging what would count as reasonable consent for those incapacitated, maybe it is the notion of reasonable consent itself that is doing all the work, and not a consent theory at all. In this chapter, I shall argue that this is a mistake. Relying on reasonable consent removes consent from the picture entirely, and with that all the support we can muster from evolutionary models. Appealing to idealized consent means the deciding factor is reason. The focus is what a reasonable agent would do, whereas evolutionary forces are not concerned about reasonable agents. As Blackburn notes, ethics involves coordination, not reason.[4] Similarly, Skyrms highlights how morality emerges from correlated conventions, not rationality.[5] Although coordinated conventions can come about in non-rational ways, consent is still key, since it is only through the correlated strategies that moral conventions arise. It is this emphasis on coordination that highlights the centrality of consent over rationality in ethics. Consent is key. For this reason, accommodating proxy or hypothetical consent is not getting us closer to the nature of morality, but further, and we tolerate it only in the special cases when one has not the cognitive capacity to give consent (or non-consent), or has not all the relevant information. Falling pianos, surprise parties, and proxy consent for the incapacitated are situations that deviate from the conditions in which morality takes root. On my model, someone's notion of idealized consent never trumps actual non-consent. To do so is to abrogate the lessons about the success of coordinated conditional strategies in evolutionary models. I accuse the extant contract theories of reversing this: placing idealized consent over actual consent.

The normative principle of contract theories rely too heavily on what is rational or reasonable. In Chapter 4, I have argued that such attempts ultimately fail. It is not reason that leads to morality, unless we mean by reason that which has evolutionary fit. Reason appeals to the agent, evolutionary fit appeals to the continued success of the gene. But since I may have no reason to care about the continued success of my genes, I will not have a reason to be moral. Contract theory must adjust itself in order to accommodate its evolutionary support. The normativity of standard contract theories focuses on what idealized agents would agree to. Of course you ought to do what your perfectly idealized self would agree to do given full information and similar circumstances, etc., but what would that be? What would perfectly idealized agents agree to do? The question suggests the answer will take a substantive form, and that is where the idealization goes astray. For evolutionary ethicists, the idealization can be given very concrete form: the strategies that have evolutionary fit are the ones agents should fashion themselves after, so long as the conditions that drive this particular strategy to dominate remains relatively stable. Evolutionary success is predicated on the adoption of conditional cooperative strategies. In terms of moral principles, this is best translated as not doing unto others without their consent, conditional, of course, on those others agreeing to that as well.

We want to distinguish normative advice from the grounds for giving normative advice. The normative advice is simply not to do unto others without their consent. The grounds for doing so is not that such restrictions will maximize your expected utility in every case, but that those who do so tend to do better than those who do not. It is a probabilistic claim and can only ever be a probabilistic claim, but it is a good probabilistic claim for all that; one that bettors should take heed. An evolutionary point of view offers sufficient support for a system of hypothetical imperatives. In any event, it is all the backing one is ever going to get. I will next explain why consent is absent in current contract theories, and then argue for a contract theory that makes consent centrally important. I call my version *consent theory*.

6.2. Contractarianism and Contractualism

Consent theory advocates that morality serves a purpose, and that is to better enable individuals with disparate aspirations to cohabit peacefully. As individuals prosper within a moral realm, as compared to an amoral realm, evolutionary forces push individuals into developing moral dispositions and to support the development of moral norms to enlist more conditional cooperators. In this sense, morality is an instrumental norm to resolve interpersonal conflicts. Although this fits with contract theories in general, nothing in this account forces us to hold that morality is in *each* person's interests. Morality serves only those individual's interests who find themselves in interpersonal conflicts under certain conditions. Nothing can guarantee those conditions are met for everyone, but the dynamics of strategic interaction are such that it is a good bet to endorse morality. The emphasis on evolutionary forces rather than rational self-interest distinguishes consent theory from other forms of contract theories.

There are two strands of contract theory, a Hobbesian strand, exemplified by Gauthier, and a Kantian strand, exemplified by Rawls and Scanlon.[6] Some call the Hobbesian view contractarianism and the Kantian view contractualism.[7] I shall follow suit. When I speak of them both together, I shall use the more general label: contract theories. The two tackle different projects. Contractarianism tries to derive morality from non-moral beginnings. Contractualism rejects this project in favour of developing a robust morality from minimal moral precepts. The first accuses the second of begging the question. The second accuses the first of committing the is-ought problem.[8] Both argue that morality best serves one's self-interest, and their arguments for that depend on our not understanding self-interest *simpliciter*. The standard contractarian/contractualist response is to speak only in terms of a subset of preferences: those that *would* be in an agent's interest under certain *hypothetical* conditions. That is, the choice that would maximize *A*'s self-interest is not necessarily the choice *A* would necessarily prefer. For Gauthier, one must first adopt a constrained maximizing disposition.[9] For Rawls, the rational social arrangement is one that someone would choose if situated behind a veil of ignorance.[10] For Scanlon, it is what choice one would make if one had an interest in securing reasonable agreement with like-minded others.[11] These are appeals to an *ex ante* choice, as Gauthier expresses it.[12] For both contractarians and contractualists, then, self-interested actions and moral actions coincide under suitable constraints.

Admittedly, the "suitable constraints" will vary widely between and within the two camps. Scanlon will demand the constraints include substantive content about one's own self-interested desires: namely those self-interested desires that aim at arriving at agreement. It is not so much what one agrees about that interests moral cooperators, but that those cooperators' substantive desires are to reach an agreement. With Rawls, suitable constraints involve veil of ignorance metaphors. For Gauthier, the suitable constraints entail one is acting to further one's considered preferences, not one's unconsidered preferences.

Despite these differences, however, from the perspective of any *ex ante* situation, being moral serves one's interests, even when moral behaviour *ex post* does not serve one's interests. In the case of contractualists, the ideal situations are determined by moral considerations of fairness and respect, whereas contractarians are adamant to keep moral presuppositions out of the picture. Building moral considerations into the framework of the *ex ante* situation begs the question, say the contractarians. But starting from a completely amoral initial position will forever be unable to get to a morality without committing the is-ought gaffe, say the contractualists. Although contractarians commit no is-ought gaffe (my defence against the is-ought gaffe in 5.3.1 applies equally well for contractarianism), they do succumb to a case of question-begging, a point on which I will elaborate in 6.3. First, however, let me describe in what way both contractualism and contractarianism eschew the role of consent.

6.2.1 Contractualism

For Kantian contractualists, humans are autonomous agents that deserve to be treated as ends-in-themselves, and this predates any social contract, hypothetical or otherwise. In order to respect autonomous agents' interests, one must seek their consent in any interaction that involves them.[13] Failure to get their consent creates an obligation to refrain from interfering with them. Securing consent, in this sense, is a sufficient condition of treating individuals with respect. It is not a necessary condition however: one may be bound by duties irrespective of one's occurrent consent. It is important to see, therefore, that the concept of consent does no justificatory work in this picture. Moral action is not determined by consent. Consensus follows from, as opposed to derives, the duty of respect.[14]

Contractualists emphasize that the outcome of a fair procedure will necessarily be just, and anything that derives from that procedure is thereby justified; and any outcome that is inconsistent with what would result from the fair procedure is evidence of an unjust distribution. In Scanlon's terms, any act is moral so long as it would be permitted by reasonable agents motivated by a desire to follow rules that no reasonable person could reject.[15] For Rawls, any act or distribution of goods is just so long as this would be agreed to by rational agents situated in appropriate hypothetical contexts.[16] The emphasis on what reasonable persons would consent to clouds the subsidiary role consent actually plays. In Rawls's case, the suitable constraints of the hypothetical bargaining position rely on a pretheoretical understanding of fair distributions. From this, the task is simply to erect a mechanism that best yields this preconceived notion of a fair distribution.[17] The mechanism for Rawls is the famous veil of ignorance. From the vantage of ignorance concerning what role one will be assigned, persons accordingly make an agreement on just distributions. But the term "agreement" here is misleading.[18] It is not a group who goes behind the veil of ignorance. One person alone suffices. Her decision will (supposedly) cohere with anyone else's decision from the same position. One decides for all. Therefore the outcome of Rawls's model is neither strategically negotiated, nor naturally implicit. The contractualist "consensus" is imposed.

Gauthier makes this same point by distinguishing parametric from strategic choice.[19] Parametric choice entails decisions made in which the outcomes are not dependent on the choices of others. Strategic choice occurs when the interdependence of each agent's strategy is integral. Strategic choices are made relative to others' choices, which, in turn, are limited by one's own choice. A game of chess is strategic, deciding whether or not to take one's umbrella is parametric. Given this distinction, the choice from behind the veil of ignorance is parametric, not strategic.[20] And consent involves strategic choice, not parametric choice.[21] In Rawls's account, agents' interests are predetermined as being identical since any heterogeneity has been, ex hypothesi, wiped bare. The consensus is predicated on the pretheoretical notion of fairness the veil of ignorance was designed to foster. Consent, in terms of strategic negotiation among parties with competing interests, has nothing to do with it.

Rather than demanding a condition of ignorance, Scanlon imputes a shared motivation to reach reasonable agreement. Like Rawls, though, the resulting "agreement" is not the result of a strategic bargain, but imposed by the motivation condition. Also like Rawls, Scanlon appeals to a hypothetical construct in which we test outcomes. In Scanlon's terms, any act is moral so long as it would be permitted by reasonable agents motivated by a desire to follow rules that no reasonable person could reject.[22] What one would agree to, assuming one is reasonable, is simply to ask what is reasonable. But what is reasonable? Part of the meaning of reasonableness on this picture is to have the motivation to abide by rules to which we can all agree.[23] That this begs the question does not bother Scanlon: that is part of his point. We have reached the bottom of the well: a self-evident axiom about which disagreement is meaningless. In this sense, Scanlon offers an unabashedly non-reductionist account akin to the early 20th century intuitionists.[24] Ignoring the debate between intuitionists and reductionists, my point here is that the motivation to seek consent predates the consent in this picture. Consent is not what determines morality. Morality is a given and it is that which determines suitably motivated persons to agree. Consent does no justificatory work. The motive cannot then be derived by consent. Unlike with Rawls, however, consensus is a possible outcome of these imputed motives. Beginning with shared motives to agree does not mean no strategic negotiation need take place, as the Dating Game (5.1.2) amply demonstrates. But concerning negotiations to have these original motives, Scanlon is not dealing with consent.

6.2.2 Contractarianism

While Scanlon appeals to an intuition he has that we are, or ought to be, motivated to seek reasonable agreement, Gauthier tries to account for such a motivation in terms of self-interest.[25] The seeming paradox of such a demand is avoided so long as we admit various types of self-interest: those which prudence endorses and those which prudence does not endorse. So long as the pursuit of some of my self-interested desires and preferences will impede or prevent some of my prudentially approved interests, I may be motivated to accept constraints on pursuing the former.[26] We are to adopt plans that afford us the most favourable opportunities even though this may require us to act in non-maximizing ways relative to our occurrent aims. The development of morality from such austere beginnings follows indirectly. To maximize my self-interest in social contexts, I must convince you that I will not defect against you in any cooperative venture we have reason to engage in together. But to do this, I need to abandon a simple maximization mode of behaviour. I am unwilling to abandon a simple maximizing mode if the net utility of such a change is worse for me than forgoing such change. Otherwise use all the "helps, and advantages of Warre."[27] Scanlon's primary motivation for reasonable consensus has no place in Hobbes's world. Contractarians argue that reasonable agreement is where we must arrive at, not start from. The disposition I adopt must be sufficiently sensitive to prevent my being exploited. To achieve this is to recognize that the choice between unilateral cooperation and unilateral defection is a false dichotomy. I might adopt a more sensitive disposition; a disposition that directly depends on the disposition of the others with whom I engage. So long as I can internalize this constrained maximizing position, I can assure myself of doing at least as well as straightforward maximization of self-interest, and so long as I can interact with others prone to cooperate, I can in fact do better. Given the social constraints in which I operate, the best way to maximize my self-interest is to adopt a disposition prone to comply only with other compliers.[28] To reap the benefit of this advantage, I must accept constraints on my temptation to resort back to straightforward maximization. Internalizing constraints is the contractarian link to morality.[29] The benefit to me from cooperation will be annulled if the other unilaterally defects, leaving me in a worse spot than had I not agreed. So in order for me to enter any agreement, I need assurance that the other will not renege. Absent such assurance, no one who is interested in maximizing self-interest will enter any agreement. Since cooperative ventures are the only way I can maximize my self-interest given the social construct into which I have been thrown, I will be motivated to pay the costs to prevent unilateral defection. This cost will entail adherence to such rules myself. If the rules are to be externalized, I will also add the administrative costs of policing such adherence – so long as all these costs are less than the gain I hope to reap from cooperation, as well as beat out the expected gain from mere internalized constraints, if such are possible. The limiting assumptions include the belief that persons equally recognize the unlikeliness of achieving their most preferred outcome and have reasons to fear yielding their worst preferred outcome. Given this, negotiating with others offers the hope of advancing one's position beyond the status quo.

The intricate tangle of strategic interaction inherent in the contractarian account is completely absent from the parametric account that contractualists offer. Kantian contractualists require the motivation to be tied to the good will, and not self-interest. To conceive morality as the *result* of a contract, however, we must conceive contract negotiation from non-moral motives. But just as contractualists did, contractarians also eschew the role of consent.

When we are assessing the justification of the outcome, we appeal to whether the procedure used was the one upon which we all agreed. When the question concerns what moral or political institutions we ought to endorse, the contractarian idea must move to a more abstract domain: we must seek a procedure for procedure-selection. And if we wish for the justification of this second-order procedure, we move to a tertiary agreement on that procedure. The only way to preclude an infinite regress is to assert that somewhere along the justificatory map we no longer appeal to what people actually agree to: instead, we appeal to what rational agents suitably situated *would* agree to in a premoral bargain. Assuming that such agents agree only to that which is reasonable, the justificatory schema is reason, not agreement. What people agree to in fact, it turns out, is not necessarily what contractarians would sanction. If x is the reasonable choice, then whether people agree to y is irrelevant. People are, after all, dim-witted. "Indeed," Gauthier admits, "I do not even suppose that the practices with which we ought willingly to comply need be those that would secure our present agreement."[30] Gauthier worries that "agreement" can come about in a whole range of inappropriate ways. One may be manipulated, lied to, coerced, brainwashed, or deceived. Alternatively, one may be incompetent. In none of these cases can we properly speak of consent or agreement. This echoes Rawls's concern: to ask whether a current arrangement was agreed to by all parties is insufficient, since their agreement is only conditional upon their current status quo, but that status quo may itself be unfair.[31] Asking what one would agree to assuming one is rational is simply to ask what is rational. Consent has nothing to do with it. Both contractarianism and contractualism are misnomers.

Summary. I have so far highlighted the extent to which reliance on consent is marginalised in both strands of consent theories. Contractualists rely on pre-theoretical moral concepts, whereas contractarians are thought not to. Self-interest alone is deemed to be the grounding for morality. But in so arguing, the concept of agreement is now thought idle. For both theories, the concept of consent does no justificatory work. Scanlon would claim A and B's consensual action is moral only so long it is an action reasonable persons seeking consensus would agree to do. So the mere fact that A and B agree is not the relative factor. Likewise for Rawls. Agreed upon actions may not be morally permissible so long as those are not the actions that would be chosen from behind the veil of ignorance. The occurrent bargainers may start under unfair advantages which contaminate the ideal procedure. Gauthier is no different in rejecting occurrent consent as a sufficient indicator of moral action. Agreement in practice may fall short of what agreement in the *ex ante* position would entail. For Gauthier, the contractarian test *idealizes* the appeal to agreement even at the normative level. It is clearly not what persons would agree to given their actual social circumstances, but what rational persons would agree to were they to choose *ex ante* their social circumstances.[32] The official doctrine of both contractualists and contractarians is that talk of "consent" at the normative level is irrelevant. In the place of consent, contractualists offer a non-reductive intuition and contractarians appeal to normative rationality. Appealing to an evolutionary stable strategy, however, brings consent back into the picture in terms of highlighting coordinating strategies in nature.

6.3. The Incoherence Problem

While Scanlon appeals to his intuition that we are motivated to seek reasonable agreement, Gauthier tries to account for such a motivation in terms of rational self-interest.[33] My position is that neither are right. I offer a more naturalized account of the motivation explained in terms of evolutionary fit. But my disagreement with finding the motivation to seek reasonable agreement in rational self-interest is different than merely the argument discussed in Chapter 4. I shall argue that the Hobbesian contractarian position meets a fundamental incoherence.

Hobbes's external solution faces the most obvious incoherence. Hobbesian agents cannot get along well without a sovereign to oversee their conflicts and settle disputes. Nevertheless, the claim is that we come to recognize the need for this sovereign, and we all voluntarily decide to give up our "rights." But how can we come to such an amicable agreement prior to the sovereign's rule? If we can come to agreement on that issue, then what need do we have of the sovereign for other agreements? On the other hand, if we require the sovereign to settle all our disputes and oversee all agreements, we could never get out of the State of Nature for there is no sovereign yet to oversee the election of the sovereign. As Hampton remarks, "Hobbes's account of conflict [in the State of Nature] seems to generate sufficient strife to make the institution of the sovereign necessary, but too much strife to make that institution possible."[34]

Gauthier's appeal to an internal solution does not avoid the problem. Agents do not consent for consent's sake. Nor is it the case that moral duty originates magically merely at the point where one says I promise.[35] This would presuppose promises are binding, but we want to know how such a thing is possible. It is a poor answer to simply point out it is because we have promised, let alone because artificial agents in an artificial world would have promised. But how is the obligation established? That I am motivated to agree to a duty does not show I am now motivated to abide by my duty. Saying I will do *x* is not normally sufficient to permit societal interference should I fail to do *x*. If my decision at the *ex ante* level were merely parametric, the solution would be at hand. I would agree to do *x* given the pre-existing social constraints and it is the continuance of these constraints that continue to motivate me to do *x*, and not merely to give lip service to my doing *x*. But when the matter concerns why we would continue to abide (and uphold) these constraints themselves, we can no longer appeal to the social constraints which bind our *ex post* actions. We need a reason to continue to uphold these constraints. Similarly, we may have a reason to obey a 60 km/h speed limit once radar traps are in place, but the mere institution of speeding tickets cannot itself justify setting the speed limit at 60 km/h. Some other argument is required for that.

At the *ex ante* position, we are agreeing on what moral principle to adopt – not for its own sake, but rather for instrumental purposes. Contractarians see moral principles as necessary stepping stones toward achieving our self-interested goals given existing social constraints. To negotiate with others, agreement on a few principles concerning the structure of negotiation itself is first required. These need not be explicitly agreed upon, any more than the conventions of language but, absent agreement on these, just as absent coordination on language, negotiation is useless. Hobbes believed one such principle was to keep one's agreements.[36] We need to know up front that consent binds us so that decisions are not made lightly. I will not agree to anything to which I would be unwilling to commit myself. A rule governing adherence to one's promises helps assure against defection, thus removing one of the obstacles to cooperation.

As Hume noted, however, if I do not agree to this "keep-my-agreement" principle to begin with, nothing precludes my "consenting to it" without in any sense being bound by that promise. If others do feel bound by this principle, better fodder for me. But given rough equality in *ex ante* negotiations, others will realize this loophole too. Hobbes's principle is empty. Of course, keeping one's agreement is a necessary means to keeping the peace (Hobbes's first Law), but Hume's worry is not eluded. That I agree to seek peace does not in any sense obligate me to seek peace when it seems in my interest to break peace. To agree to abide by the outcome of negotiated consent is clearly a prerequisite to entering negotiations in the first place. An underlying circularity exists.

Contract theorists ask what moral principle, or political structure, pre-moral rational agents suitably situated would endorse. This *ex ante* procedure expressly maintains that the outcome of such a procedure is morally binding. Therefore, it behoves contract theorists to show that the very procedure of ascertaining moral principles is what these hypothetical agents in an *ex ante* position would endorse. Of course, to endorse the rules for an *ex ante* agreement, they would need to be in an *ex ante ex ante* agreement. Consent at such an abstract level can hardly count as consent: but admitting consent is merely a heuristic ploy does not evade the problem. An appeal is made concerning what ideally rational agents would do. Supposing these hypothetical agents reach an agreement, how can such a counter-fact be considered a justification for what we are to do? It is analogous to my deciding that the rules of the game will be decided by me, including the procedure to decide the rules of the game, and proceed to justify my decisions on the basis of that procedure. And surely it is conceded that if our hypothetical agents in the *ex ante ex ante* position do not choose the contractarian procedure for moral justification, their choice could hardly count, given that it is an outcome of a procedure they reject. That is, they admit consent has nothing to do with it. When lack of consent fails to count against the proposed principle, consent to it cannot count in its favour. The contractarian project presumes a normative justification; it cannot justify it. In order to make any agreement – whether behind a veil of ignorance, or presuming shared motivations to seek agreement, or constrained by conditions of equality, rationality, amorality, non-tuism, limited resources – they would already have to endorse the moral force of their consent. So modern contract theories succumb to the same sort of logical incoherence as Hobbes's appeal to the sovereign. *Ex ante* consent grounds moral norms, but *ex ante* consent is itself deemed morally apt. Hypothetical constructs are already predicated on the viability of consent.[37] To agree on the institution of morality seems, therefore, to involve an infinite regress. As soon as our hypothetical agents consider negotiating for moral constraints, they presuppose the outcome of their agreement is morally binding: they presuppose the validity of consent. The validity of consent will forever predate their agreements about it.

Both contractarians and contractualists claim that consent is merely a heuristic device for tracking the maximization of preference satisfaction under suitable constraints. Neither camp – contra Hume – suggests that consent is binding. Rather, consent tracks what is rational, and it is its rationality that provides the normative force. This, at any rate, is the official position. But it collapses at closer inspection. For (1) if x is rational for

A to do, we would not normally say *A* is obliged to *x*. If *A* fails to do *x*, we might conclude *A* is irrational, but not immoral. And (2) we cannot conclude the moral choice is rational in cases of strategic interaction. If *B* cooperates in a prisoner's dilemma (PD), *A* would be irrational to cooperate as well, given *A*'s preferences – even *A*'s considered preferences, since considered preferences concern the substantive content of preferences, not the preference structure which may or may not lead *A* into a PD. So even if it is rational to be moral, no obligation is derivable from that. Besides, as illustrated in Chapter 4, it is not rational to be moral. In no way is Hume's worry avoided by modern contractarians' reluctance to treat consent as anything but a heuristic device – so long as they emphasize rational morality.

Abandoning the reliance on rationality removes the incoherence worry. To ask why we should agree to the principle of consent is the wrong question. Rather, the dynamics of strategic interaction presuppose the normativity of consent theory, just as the dynamics of language presuppose an evolved coordination. This is the position I shall elucidate next.

6.4. The Resurrection of Consent

The evidence from evolutionary game theory models suggests that conditional agreement has a much more central role to play in theory than either contractarianism or contractualism offers. While contract theories have emphasized that what people agree to in the occurrent, or *ex post*, world cannot override what ideal agents would agree to in the hypothetical, or *ex ante*, domain, my point here is the reverse. What people agree to in the hypothetical domain must be modified according to what people would agree to in the occurrent world; it cannot be otherwise if *ex ante* arguments are to be motivating. To help see why, let us divide the contractarian tradition along another dimension: that between Hobbes and Locke, as opposed to Hobbes and Kant. Hampton calls the Hobbesian tradition the "alienation contract model" and the Lockean tradition the "agency contract model."[38] Citizens of the agency model lend their power to the sovereign state and retain their right to revoke it. Agents of the alienation model surrender their power to the sovereign state and have not the right to revoke it. At the political level, the agency model is generally deemed preferable. We wish to retain our political autonomy – our right to revoke a particular sovereign's power, our right to protest, our right to cast a meaningful vote. At the moral level, however, the agency model is deemed inadequate. The ability to revoke a moral right cannot predate the moral institution. If the right we can revoke is understood as simply the right to renege on an agreement, then this is precisely what will upset any solution to the prisoner's dilemma (PD): we will be unable to escape the State of Nature.[39]

Hobbes's alienation model is not to be interpreted as revoking *all* of one's liberties, however, which would be insane. We certainly have the right to revoke our agreements so long as the other has revoked her part of the bargain. Agreement is thereby a condition of *unilateral* non-revokability. Given the cognizance of unilateral non-revokability of commitments, one's commitments will necessarily be restricted to only those deemed necessary. You would be unwise to commit yourself to actions you will later regret. Complete surrender of all powers is therefore excessive. When Odysseus bound himself to the mast and commanded his crew to ignore his orders to untie him and row him ashore to wallow with the Sirens, he did not concede to having all of his orders disobeyed. While listening to the Sirens, he might ask for an umbrella, or some water, or to ask that the disgruntled cabin boy stop whipping him, or the crew to stop beating their drums so loudly they obscure the voices of the Sirens. The non-revokability of his commitment must be limited to only that which is necessary. To adopt the full agency model will also not do, since Odysseus's plan would backfire. Under the full agency model, his *ex ante* command would be interpreted thus: "While sailing past the sirens, disobey my pleas to be let loose, unless I plea to be let loose." Of course he will so plead, and ultimately he will join the "piled-high mouldering skeletons of men, whose withered skin still hangs upon their bones."[40] In order for him to get the best of both worlds, Odysseus must commit himself in a way that the agency model precludes. But the pure alienation model is also senseless. His being bound must remain conditional and limited.

Since our *ex ante* commitment limits the degree of our liberty, we must not ignore this in our precommitment strategy. A rational act entails not paying more than one needs to. As Odysseus did not want his crew to disobey *all* his potential commands, *ex ante* agents will be loath to commit themselves to points beyond their self-interested motivations. Similarly, we are loath to admit that any agreement is necessarily moral, for involuntary agreements, incompetent consent, and negative externalities must be precluded to elicit *ex ante* consent. Curtailing consent need not show that consent's role is marginalised. Rather, the concept of consent itself stipulates these constraints. The concept of "involuntary consent," for example, is analogous to a cancelled concert.[41] A cancelled concert is not itself a concert. It might have been a concert had a particular thing not occurred (the cancellation). Similarly, an involuntary agreement is not itself an agreement. It might have been an agreement had a particular thing not occurred (the involuntariness). The very concept of agreement and consent, then, entails voluntary choice. Likewise, having appropriate information is part of the very meaning of consent; it is not an anterior constraint. We cannot consent to terms of which we are unaware. And to be aware of information and to be subject to voluntary action both require competency. The incompetent can neither process information nor make truly voluntary actions.[42] Finally, assuming concerned parties are minimally identified as those adversely affected by the issue of the agreement, the condition that *all* concerned parties voluntarily agree rules out the prospect of negative externalities.[43] (I discuss this condition more fully in Chapter 7.) When we say consent or agreement, then, we mean it in this qualified sense. Consent and *proper* consent are identical. Improper consent is not consent at all. "Consent" refers to voluntary agreement among all suitably informed, competent concerned parties.

Consent, so defined, is not merely a heuristic device to describe strategic solutions of ideally rational agents. Consent in my version of contractarianism plays a much more central role. In evolutionary terms, genes that build vehicles that have – among other useful phenotypic traits – the strategy to conditionally abide by one's agreements will be the ones that take over the largest proportion of the population. Rationality is not the driving force; the ability to conditionally keep one's promises is.

Ruling out coerced agreement by definition may strike one as cheating. The principle of consent is supposed to define morality, but, as contractarians accuse contractualists, moral concepts (in this case strictures against coercive agreements) are embedded in the principle of consent itself. A defence of this manoeuvre is in order. First off, consent is not the moral issue: keeping one's consent is. Understanding consent as necessarily voluntary does not magically impel one to keep one's promises. Recall my account of the prisoner's dilemma (PD) in Chapter 4. I described the PD in terms of a contract between a bootseller and someone wishing to buy boots. I called the idea of the exchange the contract. The contract merely defines the mutually beneficial outcome. The mutually beneficial outcome is stipulated by the voluntary preferences of the agents considering the contract. For mutual benefit, coercion is ruled out a priori. It is possible one may be coerced into doing that which one prefers, but this would not be the norm: coercion is useful precisely to get people to do what they do not prefer. The contract devolves from the real preferences of the players. It is not imposed on them from without. For example, someone not in the market for boots cannot be said to have entered the contract with the bootseller. Someone not interested in selling boots cannot be said to have entered the contract with a boot seeker. The contract just is the mutual optimal outcome of the voluntary preferences of players in negotiation. Those who enter the contract do so voluntarily.

To voluntarily enter the contract is to voluntarily accept the sanction against coercion. To voluntarily enter the contract, so defined, however, is not to thereby be morally bound to honour the contract. That is a different matter, and that is the matter with which moral norms deal. Therefore, consent's coming fully loaded to ward off coercive agreements is not what is doing the moral work. Strategic negotiation itself presupposes the thick concept of consent. This is not to mean everyone has a moral obligation in the state of nature to make agreements. No one has. Only those who stand to benefit from interaction with others are moved to make contracts, and consent to those contracts must be understood as uncoerced or voluntary. The contract merely entails the structural arrangement of preferences under strategic negotiation. The moral issue is whether or not one reneges on that contract. Unlike Scanlon's admission, consent theory is not based on a moral intuition for which there is no deeper justification. Rather, the very concept of a hypothetical choice situation already presumes the normative force of the consent principle. In the following section, I shall back this bold claim in game theoretic terms. But first, a summary of the argument may be in order.

Summary. In 6.2, I argued that both contractarianism and contractualism eschew a reliance on consent. In 6.3, I argued that they were mistaken for doing so, and in 6.4 I argued that consent is a crucial ingredient in understanding moral theory. Moral constraint against coerced agreements cannot be derived from any *ex ante* agreement, for the very concept of agreement must presuppose it. Thus, even the *ex ante* agreement must already presume a more basic principle of consent: namely *any act is moral only so long as all concerned, suitably informed, competent agents agree.* This is the normative principle one can derive from examining the results of evolutionary game theory. Evolutionary modelling shows that successful strategies across disparate games are (1) conditional strategies that (2) can do well if correlated with its own kind. In more common moral principle talk, this gets translated into the principle of consent. The principle of consent is the normative principle that any contract theory must already presuppose when discussing *ex ante* agreements. We admit it cannot be rationally justified in the way that contractarians demand, but such an admission is costless, since rationality in the sense demanded is not what moves us.

Voluntary actions are those that are not coerced, nor unduly influenced, and are made by agents who are not misled, nor under false expectations, who are competent and sufficiently informed. We may legitimately wonder whether such conditions are met in practice, but meaningful discussion at that level must presuppose truly voluntary action among all concerned parties is morally permissible. And we have now seen why this is so: Without such an implicit agreement, no agreements can even be proposed, let alone kept. This point is key. In the following section, I offer a game theoretic demonstration of how the principle of consent follows from – as opposed to underwrites – strategic interaction.

6.5. Moral Standing

A consent theory may well be read as suggesting no one can have any moral standing until they are party to the social contract. This is literally true, but is nowhere near as loathsome as it sounds. Before I explain why it is not so loathsome, I will make explicit how atrocious it does sound. Those who are outside the realm of bargaining have no claim rights at all, so cannot count as concerned parties. Any holdouts remain in the state of nature where they are not protected by any rights and no one has any duties toward them. Such a literal interpretation of consent theory counteracts the moral intuition that it is wrong (or prima facie wrong[44]) to harm others without their consent. At the normative level, certainly, it is absurd to believe that only those who agree to an action have moral protection. If so, we could not complain about a gang of thugs assaulting a lone pedestrian. That the pedestrian is outside the thugs' agreement hardly shows she has no right to complain. That she was not party to the agreement does not rule out the possibility that she *ought* to have been party to the agreement. I do not mean that she ought to have agreed to the assault, but that her non-consent morally matters. Hobbes believes that mutual refraining from violence is the sine qua non of morals. Consent theory agrees. The conceptual difficulty is to explain how the notion of mutual refraining can evolve prior to the moral notion of "Keep one's promises!" That is what I shall demonstrate in this section.

Let us first distinguish two cases. In one scenario, we shall imagine someone is outside the *ex ante* agreement and is doing something bad to those of us who have accepted the *ex ante* agreement. The conditionality clause of our agreement permits our retaliation to such individuals. Our self-preservation tells us what to do with such agents and it is that fact that stands to modify their behaviour. Hypothetical moral imperatives do not recommend unconditional acquiescence. In another scenario, however, we shall imagine someone who is outside the agreement but is not doing anything bad to us. This agent is simply minding her own business. To harm her does seem to be a violation of the agreement that binds us: not to do to others without their consent. But are these two cases really different? To be "outside the *ex ante* agreement" is not to agree to peace; not to agree to our refraining from taking from them by force. If they are not planning on doing anything to us, if they really desire to mind their own business, why would they not have agreed to our *ex ante* proposal concerning the principle of consent? Plainly, to be "outside the agreement" means not being committed not to exploit others without their consent. Although rejecting such a commitment does not mean one will exploit others, it does permit others to exploit them without their consent – even in cases where they were merely minding their own business. It is precisely this worry that makes it an evolutionarily unstable strategy to reject the *ex ante* agreement on the principle of consent.

A glitch in this reasoning is thought to be exposed by examining the Prisoner's Dilemma (PD). Even though we would each do better by mutual agreement, the temptation to unilaterally defect remains too great for rational, pre-moral agents. As Hume would say, the mere fact of our *ex ante* agreement cannot bind us. I would like now to point out how this is not so. When examining PDs, we operationally define immorality as defection from one's agreement. Defection must be understood *post* agreement. Let us therefore distinguish the consenting (or the failing to consent) and the acting (or the failing to act) on that consent. In PDs, we assume all parties have consented to the terms, and we examine the logic in abiding by those agreements made. Parties make the initial agreements according to their subjective preferences. If one does not stand to gain by mutually abiding by the terms of one's agreements, one is not motivated to offer the initial consent. As a result, we have four possible outcomes from the situation where agents accept a proposal to cooperate: coop,coop; coop,defect; defect,coop; and defect,defect. Since one may also decline the initial proposal to cooperate, there are actually eight possible outcomes to worry about,[45] as indicated by the Table 1 below.

The intent of Table 1 is to highlight the difference between rows 1 and 2. Row 1 (the row marked "1. propose,accept") represents, albeit in non-standard fashion, the PD. Here, "acts" means doing what one has accepted or proposed. In Case 1, therefore, "acts" can be understood as "coop" and "doesn't" can be understood as "defect." In Case 1*a*, "acts, acts" refers to Player 1 cooperating while Player 2 also cooperates. In 1*b*, the first player

Table 1. Proposal game structure

Consent	Act			
	a	b	c	d
1. propose,accept	acts,acts	acts,doesn't	doesn't,acts	doesn't,doesn't
2. propose,reject	acts,acts	acts,doesn't	doesn't,acts	doesn't,doesn't

cooperates while the second defects. In Case 1, therefore, *a* is the moral option, *d* is the amoral option (the State of Nature), and *b* and *c* are immoral actions. But what is going on in Case 2? Here no initial agreement has been made at all. Can one technically defect from a non-agreement? In this case, doing what one has proposed despite rejection of the proposal (Case 2*b*) is what we are interested in. Normally we understand this as immoral. Consider a dialogue between Cain and Abel. Cain: "May I kill you?" Abel: "No." But if Cain kills Abel anyway, we are not inclined to say the action is moral, let alone amoral. We are prone to call the action immoral. (If the offer is accepted, and both parties act on it (as in Case 1*a*), this should count as moral, a case I shall discuss in 6.6.) But the standard analysis of the PD cannot capture this. In Case 2, the moral action is *d*. Immoral actions would occur in cases *a*, *b*, and *c*, although presumably only Case *b* is the live option. But how can consent theorists call 2*b* immoral?

Hume's complaint, by the way, concerns how contract theorists can call 1*b* or 1*c* immoral. Usually we say it is immoral given their commitment at the first stage, but Hume asks how the first stage morally binds one. We seem to need a pre-agreement that so long as one consents to cooperate, one must in fact cooperate. But our agreeing to that presupposes the very thing on which we are agreeing. Here the issue is similar. We suppose *A*'s failure to get *B*'s consent to action *x* (where *B* is a concerned party in *x*), creates an obligation on *A* to refrain from doing *x*. But how so? If morality is by consent, and *A* does not consent to this rule, how is *A* obligated? (Hume worried how *B*'s consent to oblige *A*, obliges *B*.)

The response is to note that although there is no obligation, *A* is not acting in isolation. *A*'s moves are strategic, not parametric. To illustrate, consider the tree in Figure 1.

The tree in Figure 1, highlights how other players' responses to *X*'s choice will impact the payoff to *X*. Agents' moves are conditional, not absolute, and that conditionality is key. *X* makes a choice from the initial node A. *X* may choose either B or C. If *X* chooses B, the game is over, and the payoffs are 0 for each.[46] (The left utility is *X*'s score, the right utility is *Y*'s score in each of the bracketed payoffs.) Should *X* choose C, it is *Y*'s turn to make a choice: either D or E. From the node D, *X* chooses either F or G, at which point the game is over. From node E, *X* may choose either H or I. The choice from the nodes H and I is *Y*'s to make.

Using Zermelo's backward induction, at node I, *Y* will defect rather than coop (outcome L), yielding −1 to *X*, and 2 to *Y*. *Y* so chooses since 2 is preferable to 1, which *Y* would have received had she chosen outcome M. *Y* will defect at node H as well, but in that case, the payoff is 0 for both *X* and *Y*. Since *X* knows this and deems 0 preferable to −1, *X* will choose H from

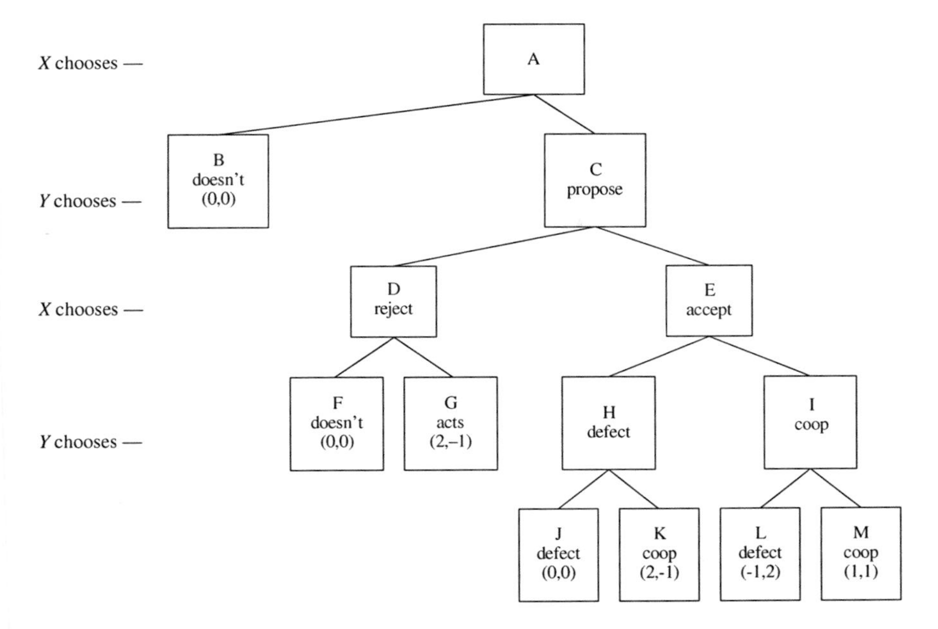

Figure 1. Proposal game tree

node E. That is, should *X* be at E, *X* will defect, and the resultant payoff is 0,0. If, on the other hand, *X* finds herself at node D, *X* will choose G (acting in accordance with the proposal despite *Y*'s refusal of the proposal) yielding 2 for *X*, and −1 for *Y*. Had *X* adhered to *Y*'s refusal, *X* would only have received 0 (outcome F.) Thus, from the choice point C, *Y* stands to receive −1 by choosing D, but 0 from choosing E. Therefore, *Y* will choose E. That is, should C be reached, the payoff for both will be 0 (outcome J). But since 0 is the payoff for both should they reach outcome B, there seems no reason to bother making any proposition at all. In fact, so long as the act of propositioning someone has costs (time and energy and perhaps risk of preemptive strikes), B will be the rational choice.

The point of the above is to show that in order to even propose a collaborative venture, one must already presuppose the binding force of contracts: otherwise there would be little point in the initial proposal. If so, Hume's question is backwards. It is not to ask how to agree on the convention of agreement without presupposing the convention of agreement; it is to ask how one could refuse the convention of agreement given one's strategic preferences. What is right about the literal rendition of the Hobbesian approach is the tautology that prior to the institution of morality, no morality exists. This illuminates how morality comes about by agreement on normative principles, and how the very concept of agreeing on moral principles already presupposes the moral force of consent, a consent to mutually refrain from exploitation and to stick with that agreement. It is therefore misguided to speak of people ever being outside the agreement at the *ex ante* position.[47] What this also shows is that the principle of consent is already presupposed at the instance of any strategic negotiation. The moral advice is to follow through with one's agreements. One's agreement is not defined in some abstract sense, but follows from whatever occurrent preferences motivated the agent to enter the strategic negotiation in the first place. In other words, should there be two groups of agents – those who seek to maximize preferences in terms of modular rationality and those prone to understand strategic negotiation as itself demanding conditional constraints against coercion, the latter will prevail. A naturalized account is all that is required.

Summary. Hypothetical contractarian credos take the following form: x is moral so long as x is an action that suitably defined agents in suitably defined circumstances would endorse. This presupposes (so I claim) the following first principle for *ex ante* negotiators: x is moral so long as all concerned, suitably informed, competent agents agree. This principle governs not merely the initial *ex ante* agreements, but all occurrent agreements. Although *ex ante* negotiators might want to agree on this as a normative principle governing occurrent social interactions,[48] it is itself the grounding of all agreements sought by self-interested agents. It is already forced upon them as soon as they devise the idea of an *ex ante* justification. It is not a matter of rational choice. Rather, the concept is a well enmeshed phenotype: it has evolutionary fit. The root of contract theory is *not* that agents consent to moral constraints, but that in order for mutual benefit to arise from strategic negotiations, they must presuppose the binding force of consent.

6.6. A Counter Case?

If we find a case where consent theory says, "Moral," while our intuitions say, "Immoral," we have a prima facie reason to doubt consent theory. William Gass offers such a case.[49]

> Imagine I approach a stranger on the street and say to him, "If you please, sir, I desire to perform an experiment with your aid." The stranger is obliging, and I lead him away. In a dark place conveniently by, I strike his head with the broad of an axe and cart him home. I place him, buttered and trussed, in an ample electric oven. The thermostat reads 450

> F. Thereupon I go off to play poker with friends and forget all about the obliging stranger in the stove. When I return, I realize I have overbaked my specimen, and the experiment, alas, is ruined.[50]

He continues, "Any ethic that does not roundly condemn my action is vicious."[51] Now, since the stranger consented to participate in the experiment, and anything that meets mutual consent is morally permissible according to consent theory, it would appear that consent theory is therefore a vicious ethic, if an ethic at all.

One could complain that the obliging stranger did not, presumably, consent to being hit on the head and burnt in an oven. He consented to partake in some experiment, and those who consent to participate in experiments do not usually assume they will be adversely affected in the process. So long as consent theorists speak of "informed consent" in grounding the moral permissibility of an act, consent theory would not be shown to be vicious after all, since the onus was on Gass to explain to the stranger what the experiment constituted. In other words, as put, reductionists can cry "immoral" just as easily as non-reductionists.

To avoid the side-stepping response above, we need to alter Gass's case. What if Mr. Gourmand approaches a stranger and asks permission to chop her up and roast her and she, knowing full well the implication of the request, obliges?[52] According to consent theory this would evidently be morally permissible, but to normal people such an act is morally atrocious, whether it was agreed to or not. So much the worse for consent theory. Why would anyone defend such a theory?

The answer is this. The cost of making this case transparently clear is its removal from the rather more muddy reality. I admit that under such a scenario, consent theorists would have no moral right to intercede. Roasting the obliging stranger is a morally permissible event. What it is not, however, is an instance that can properly capture our moral intuitions. The moral outrage we feel is based on our inability to grasp one of the requisite premises of the thought experiment. The scenario demands that we assume the stranger obliges. This is what we cannot fathom. Our intuitions dictate that the stranger would not acquiesce to the horrible request. Being unable to dismiss this normal intuition, we also feel the moral outrage in the continuance of the act. But so long as she does not oblige, or is not in a fit frame of mind to appreciate the request, our moral outrage is justified under consent theory. Even if we cannot fathom the obliging stranger's competency, this would not reveal prior moral constraints on agreements. Consent theory demands agreements to be among competent agents. Normally we understand competency as the default. Reasons for removing our ascription of competency to someone will involve, minimally, that person's inability to comprehend the logical consequences of the agreement. If a decision to be eaten for no reason necessarily counts as such a case, then, again, our moral outrage against this "agreement" is perfectly consistent with consent theory.

I modified Gass's counter case to reductionist ethics to help highlight how the normative principle of consent naturally arises from the structure of strategic negotiation itself. Morality does not advise us about what negotiations we must enter, only what to do once we enter strategic negotiations. Moral advice, meanwhile, falls from what has evolutionary fit given the social dynamics of strategic negotiation. This is why we cannot fault an agent who properly consents to be eaten, although we can complain about eating someone who does not so agree.

6.7. Summary

Contract theories suffer from an incoherence problem. Modern solutions move to a hypothetical domain: people would agree to abide by morality *ex post* once they were suitably situated in an *ex ante* position. Such manoeuvres do not highlight consent, however. Any decision at the *ex ante* level is purely parametric. But evolutionary game theory shows that moral behaviour evolves as a solution to strategic interaction, not parametric choice. Besides that, any *ex ante* decision must already presuppose the normative force of consent. Contractarian analyses portray agents bargaining on principles of justice and these principles are justified by the fact that this is what rational agents would endorse in suitable circumstances. But have these idealized agents agreed on the procedure that moral matters will be determined by the principle of consent? If not, is the resultant agreement useless? And even if they had, would such an agreement be coherent? After all, to agree on this, agents would have to presuppose that consent is binding: the very thing on which they are supposedly agreeing. That is, the doctrine of consent needs to be itself already presumed as the linchpin of morality in order for the *ex ante* machinery to get off the ground. This is not explained in terms of moral intuitions. Rather, so I argue, it follows as part of the evolution of fit strategies. Evolutionarily fit strategies across disparate games are (1) conditional strategies that (2) can do well if correlated with their own kind. This gets translated into the principle of consent. Like Hume's reliance on our senses, the principle of consent is not itself open to justification, for any justification of any contract theory already presupposes the principle of consent. The doctrine of consent may be put thusly: *Any act is morally permissible if and only if all competent, suitably informed, concerned parties voluntarily, or can reasonably be predicted to voluntarily, consent.* Put negatively, *Any act by people that negatively affects others who have not voluntarily agreed to being so affected is an immoral act.* Loosely, this may be abbreviated to the following pocket principle: *Don't do to others without their consent.*

Notes

[1] Gregory Kavka, "A Reconciliation Project," in David Copp and David Zimmerman (eds.) *Morality, Reason and Truth* (Lanham, Md.: Rowman and Allenheld, 1984), 299.
[2] David Gauthier, *Morals by Agreement* (Oxford: Oxford University Press, 1986), 157–89.
[3] *Morality, Reason and Truth*, Kavka, 305.
[4] Simon Blackburn, *Ruling Passions* (Oxford: Clarendon Press, 1998), 69.
[5] Brian Skyrms, *Evolution of the Social Contract* (Cambridge: Cambridge University Press, 1996), 63–79. Brian Skyrms, *The Stag Hunt and the Evolution of Social Structure* (Cambridge: Cambridge University Press, 2004), 23–5.
[6] See Jean Hampton, "Two Faces of Contractarian Thought," in Peter Vallentyne (ed.) *Contractarianism and Rational Choice* (Cambridge: Cambridge University Press, 1991), 36. Will Kymlicka, "The Social Contract Tradition," in Peter Singer (ed.) *A Companion to Ethics* (Oxford: Blackwell Publishers, 1991), 195–6. Gary Watson, "Some Considerations in Favor of Contractualism," in Jules Coleman and Christopher Morris (eds.) *Rational Commitment of Social Justice: Essays for Gregory Kavka* (Cambridge: Cambridge University Press 1998), 173–4. Jody Kraus, *The Limits of Hobbesian Contractarianism* (Cambridge: Cambridge University Press, 1993), 27–8.
[7] T. M. Scanlon, "Contractualism and Utilitarianism," in A. Sen and B. Williams (eds.) *Utilitarianism and Beyond* (Cambridge: Cambridge University Press, 1982), 103–28. Scanlon links his contractualist tradition to Rousseau in T. M. Scanlon, *What We Owe to Each Other* (Cambridge, Mass.: Belknap, Harvard University Press. 1998), 5. Williams follows suit in Bernard Williams, *Ethics and the Limits of Philosophy* (Cambridge, Mass.: Harvard University Press, 1985), 75. This is not uniform. Watson, for example, treats the terms synonymously. "Contractualism comes in both a Hobbesian and Kantian form" (Watson, 173), although he proposes the term "consentualism" for the Kantian variety (Watson, 183, nt. 26), and, I guess, contractualism for the Hobbesian school.
[8] Jody S. Kraus, *The Limits of Hobbesian Contractarianism* (Cambridge: Cambridge University Press, 1993), 31. This is not to rule out other kinds of complaints each sides tosses to the other. For example, contractualists also accuse contractarians of offering a purely self-regarding account of practical reason. That is, contractarians, but not contractualists, refuse to give any credence to the intuition that other persons or their well-being can provide a basic reason for one to act. For example, "our moral intuitions push us to assent to the idea that one owes that person respectful treatment simply in virtue of the fact that he or she is a *person*." Jean Hampton, "Two Faces of Contractarian Thought," 49. I discuss this objection more explicitly in Chapter 7.
[9] Gauthier, *Morals by Agreement*, 157–89.
[10] John Rawls, *A Theory of Justice* (Cambridge: Mass.: Harvard University Press, 1971), 136–42.
[11] T. M. Scanlon, *What We Owe to Each Other*, 153, T. M. Scanlon, "Contractualism and Utilitarianism," 110.
[12] *Ex ante* agreements are hypothetical and "suppose a pre-moral context for the adoption of moral rules and practices." Gauthier, 9. When we speak of *ex ante* agreements, we need to define the circumstances of such negotiations. Wide variations exist on this in the literature, but minimally we are dealing with amoral, rational, self-interested agents in a world of competition and limited resources deciding on moral principles or principles of justice to bind them in future. Variations occur in defining these terms, of course. By amoral, I do not mean *im*moral. Nothing precludes amoral agents from behaving in ways similar to moral agents in specific cases. Likewise, by self-interested, I do not mean narrow egoism.

Self-interested agents may well take care of others. Simply, nothing dictates (nor can we predict) that such caring extends to more remote relations. By self-interest, then, I mean interests or preferences or ends the agent endorses or would presently endorse, whether or not she "ought" to endorse them by some objective standard. By rational, I mean that agents are motivated to pursue the course of action that has the highest net expected utility (taking into account approximate costs and probabilities). Utility in this case is determined according to self-interest. Occurrent agreements concern agreements made *ex post*: after the pre-moral hypothetical bargain. *Ex post* decisions concern compliance with *ex ante* agreements amidst all the vagaries of reality: unequal bargaining powers, and insufficient and uncertain information.

[13] This follows from Kant's emphasis on autonomy. It is made explicit in Kant's "On the Common Saying: 'This May Be True in Theory, but it Does Not Apply in Practice'," in *Kant: Political Writings*, H. Reiss (ed.) (Cambridge: Cambridge University Press, 1970), 63, 79.

[14] See Gauthier, *Morals by Agreement*, 6, Hampton, "Two Faces of Contractarian Thought," 50–1, and Kraus, *The Limits of Hobbesian Contractarianism*, 33.

[15] Scanlon, *What We Owe to Each Other*, 153.

[16] John Rawls, *Theory of Justice*, 17–22, 136–50.

[17] This is too simplistic, as put, given Rawls's endorsement of reflective equilibrium (Rawls, 20). We have a mechanism to yield a preconceived notion of fairness. If the mechanism fails to do so, we tinker with it until it does. Still, the preconceived notion is open for critical reflection through this process as well, or so we are led to believe.

[18] See Ronald Dworkin, "The Original Position," in N. Daniels (ed.) *Reading Rawls* (New York: Basic Books, 1976), 17–8.

[19] Gauthier, *Morals by Agreement*, 21.

[20] For a counter claim, see Anthony Laden, "Games, Fairness, and Rawls's *A Theory of Justice*," *Philosophy and Public Affairs* 20 (1991): 189–222.

[21] This also conforms to Raz's account of how consent must be "performative." Joseph Raz, *The Morality of Freedom* (Oxford: Oxford University Press, 1988), 81.

[22] Scanlon, *What We Owe to Each Other*, 153.

[23] See Jean Hampton, "Contract and Consent," in R. Goodin and P. Pettit (eds.) *A Companion to Contemporary Political Philosophy* (Oxford: Blackwell Publishers, 1993), 388.

[24] G. E. Moore, *Principia Ethica* (Cambridge: Cambridge University Press, 1903). H. A. Pritchard, "Does Moral Philosophy Rest on a Mistake?" in *Moral Obligation* (Oxford: Clarendon Press, 1949), 1–17. W. D. Ross, *The Right and the Good* (Oxford: Clarendon Press, 1930).

[25] Gauthier, *Morals by Agreement*, 4. See also Rawls, *Justice*, 16.

[26] David Gauthier, "Rational Constraint: Some Last Words," in Peter Vallentyne (ed.) *Contractarianism and Rational Choice* (Cambridge: Cambridge University Press, 1991), 327–8. See also Richard Brandt, *A Theory of the Good and the Right* (Amherst, N.Y.: Prometheus Books, 1998), 110–5, and David Griffin, *Well Being: Its Meaning, Measurement and Moral Importance* (Oxford: Oxford University Press, 1986), 10–7, for similar discussions.

[27] Thomas Hobbes, *The Leviathan* (Buffalo: Prometheus Books, [1651] 1988), ch. 14, 67.

[28] Perhaps this should read "... other *reciprocal* compliers." See Peter Danielson, *Artificial Morality: Virtuous Robots for Virtual Games* (London: Routledge, 1992), 88–90.

[29] Ken Binmore misrepresents Gauthier as following Locke's natural law justification for morality in Ken Binmore, *Game Theory and the Social Contract: Vol. 1: Playing Fair* (Cambridge, Mass.: MIT Press, 1998), 14. Gauthier's use of the Lockean proviso has nothing to do with that (Gauthier, *Morals by Agreement*, 222). On the other hand, the content of an appropriate agreement for Gauthier is tied with the antecedent constraints of the minimax relative concession, and it is not at all obvious that agreements by rational persons would necessarily invoke such demands without merely presuming equal starting positions.

[30] David Gauthier, "Why Contractarianism?" in Peter Vallentyne (ed.) *Contractarianism and Rational Choice* (Cambridge: Cambridge University Press, 1991), 25.

[31] Rawls, *Justice*, 7.

[32] Gauthier, "Rational Constraint," 330.

[33] Gauthier, *Morals by Agreement*, 4. See also Rawls, *Justice*, 16.
[34] Jean Hampton, *Hobbes and the Social Contract Tradition* (Cambridge: Cambridge University Press, 1988), 136.
[35] "[W]e are not surely bound to keep our word because we have given our word to keep it" (David Hume, *An Enquiry Concerning the Principles of Morals* (Oxford: Oxford University Press, L.A. Selby-Bigge and P.H. Nidditch (eds.) [1751] 1989), appendix III, §257, 306). See also David Hume, "Of the Original Contract," [1748] in *Essays: Moral Political and Literary*, Eugene Miller (ed.) (Indianapolis: LibertyClassics, [1748] 1987), 490. And "a promise wou'd not be intelligible, before human convention had establish'd it; and that even if it were intelligible, it would not be attended with any moral obligation" (David Hume, *A Treatise of Human Nature*, L. A. Selby-Bigge and P. H. Nidditch (eds.) (Oxford: Oxford University Press, [1740] 1978), bk.III, pt.2, sect.5, 516).
[36] Hobbes, *Leviathan*, ch. 15, 74. This is his third law of nature. The first is to seek peace. The second is to lay down one's liberty (Hobbes, ch. 14, 67).
[37] For a similar argument (with a wildly different conclusion), see J. Harsanyi, "Review of Gauthier's 'Morals by Agreement'," *Economics and Philosophy* 3 (1987): 339–43.
[38] Hampton, *Hobbes and the Social Contract Tradition*, ch. 5. Hampton, "Contract and Consent," 380. Morris refers to this distinction as the *mutual advantage view* (for the alienation model) and the *consensualist model* (for the agency model) in Christopher Morris, *An Essay on the Modern State* (Cambridge: Cambridge University Press, 1998), 7. Elsewhere, he suggests the divides should be named "agreement as consent" and "agreement as advantage." See Christopher Morris, "A Contractarian Account of Moral Justification," in Walter Sinnott-Armstrong and Mark Timmons (eds.) *Moral Knowledge? New Essays in Moral Epistemology* (Oxford: Oxford University Press, 1996), 219.
[39] This is the gist of Hobbes's infinite regress argument. See Hampton, *Hobbes and the Social Contract*, 98–105. See also Kant's argument in "On the Common Saying," 81.
[40] Homer, *The Odyssey* (Hammondsworth: Penguin, 1973), Book XII.
[41] They are both *alienans* adjectives. P. T. Geach, "Good and Evil," *Analysis* 17 (1956): 33–42.
[42] They are like Frankfurt's "wantons." Harry Frankfurt, "Freedom of the Will and the Concept of a Person," in H. Frankfurt, *The Importance of What We Care About* (Cambridge: Cambridge University Press, 1998), 11–25. Originally published in *The Journal of Philosophy* 68/1 (1971).
[43] Why would amoral agents agree to this? Because otherwise others may do to them without their consent. Gauthier's evocation of the Lockean proviso is to do the same (Gauthier, *Morals by Agreement*, 200–5).
[44] I insert "*prima facie*" here to placate those who believe individual rights may be trumped by the greater good of humanity. My real view is to leave it out.
[45] Technically there are sixteen, but since we are tracking whether one acts on the proposed plan, we assume a proposed plan is a necessary precursor. Therefore the prospects of two "rejections" of a proposed plan is simply not to have uttered a proposal at all, in which case "acting" on a non-proposed plan is incoherent. And for simplifying purposes, who proposes and who rejects will not matter, thus "propose,reject" will be equivalent to "reject,propose."
[46] The scoring represents the reverse ordinal ranking adjusted to zero. One's first choice is to unilaterally defect, one's second choice is mutual cooperation, one's third choice is mutual defection, and one's fourth choice (of four options) is unilateral cooperation. I have reversed those rankings so that the higher number represents higher utility satisfaction, defect,coop = 4,1; coop,coop = 3,3; defect,defect = 2,2; and coop,defect = 1,4. Since mutual defection is one's status quo, this outcome is adjusted to zero. Hence, the following payoffs result: defect,coop = 2,−1; coop,coop = 1,1; defect,defect = 0,0; and coop,defect = −1,2.
[47] In *Morals by Agreement*, Gauthier admits a pre-bargain rule against coercion (192) and exploitation (205), and he surmises that no rational agent would seriously enter negotiations without such precommitment rules (225). I concur. My emphasis here is that these pre-commitment rules are already embedded in the very concept of occurrent consent.
[48] Harsanyi would agree so long as we define a concerned party as one who stands to benefit from any possible interactions among any agents (J. Harsanyi, "Morality and the Theory of Rationality," in A. Sen and B. Williams (eds.) *Utilitarianism and Beyond* (Cambridge: Cambridge University Press, 1982),

44–6, 56–60). That is, occurrent agreements may be curtailed in favour of utilitarian principles. I side with Gauthier in limiting concerned parties to those who are made worse off, not merely not benefited (Gauthier, *Morals by Agreement*, 203–4).

[49] William Gass, "The Case of the Obliging Stranger," *The Philosophical Review*, 66 (1957): 193–204.

[50] Gass, *The Philosophical Review*, 193. Presumably we would not express less outrage if the specimen was properly cooked, as opposed to being overbaked.

[51] Gass, *The Philosophical Review*, 193.

[52] I owe this rendition to Louis Groarke in conversation.

Chapter 7

CONCERNED PARTIES

7.1. The Problem

7.1.1 Introduction

In the previous chapter, I left you with the following principle of consent: *Any act is morally permissible if and only if all competent, suitably informed, concerned parties voluntarily, or would voluntarily, consent.* I tried to defend it in terms of naturalized ethics and argued that consent theory captures what evolutionary game theory results would look like in terms of normative advice. In this and the following chapter, I shall defend consent theory against two practical problems that may be raised against it. In the next chapter, I shall look at the problem of charity. If two people agree to row a boat, and this action does not adversely effect others not party to the agreement, nothing could be immoral about the act, or so consent theorists would avow. Many strongly protest. It is often supposed, for example, that one's moral duty is to give positive aid to the suffering. Two people who consent to row a boat may then be immoral, according to these objectors, if a third were drowning and they were rowing the boat away from the victim. Morality, it is claimed, must require more than mere agreements. Morality must impose strictures on the content of particular agreements and, failing that, any contract theory is "morally monstrous."[1] The puzzle I plan to address in this chapter, however, concerns the concept of a concerned party.

The merit of the principle of consent rests on how we delimit the concept of a concerned party. This is not consent theory's only conceptual worry. We also need to define "voluntary" and "competency" and "informed," but there is already much literature on those concepts. An ambiguity lurking in our understanding of "consent" will be discussed in Section 7.4. If we define a concerned party too narrowly, we may be unable legitimately to prevent others from doing things that adversely affect us. If we define a concerned party too broadly, others will have veto rights on actions that ought to concern ourselves alone.

7.1.2 Pretheoretical Intuitions

Above, I spoke of actions that *ought* to concern ourselves alone. In doing so, it will appear that I am seeking a criterion to match my pretheoretical intuitions. Readers might be bothered with such an approach, even if they agree with my intuitions. Consider homosexuality. If we accept both the principle of consent and that homosexuality is morally permissible, we must exclude busybody homophobes from being "concerned parties."[2] If we count those opposed to homosexuality as concerned parties, then homosexuality is immoral, since a concerned party forbids it. Is the homophobe a concerned party? I would like to say that homophobes are not properly speaking concerned parties to the actions of homosexuals, but the homophobes say differently. How can I be sure my concept of a concerned party is not unduly biased toward my pretheoretical views on homosexuality? Similarly, the following sort of bold statements will appear below: "Surely targets of racist and sexist jokes count as concerned parties," and "Surely persons suffering from second-hand smoke count as concerned parties." That my final definition satisfies my intuitions on these matters does not show my definition to be superior, since it is logically possible that my intuitions are corrupt. In my defence I shall assert merely that the burden of proof lies with serious objectors to these claims. Those who assert that homophobes are concerned parties to the agreements among homosexuals, or that members of groups targeted by racial or sexist slurs are not concerned parties, need to provide an alternative criterion that we can assess against a backdrop of moral judgements we do share. My task, recall, is to carve out normative advice that best matches the results from evolutionary games, not to presuppose the normative advice is true, or rational. I raise the problem of bias partly to make explicit that an evolutionary account of morality cannot get mired in talk of what content of agreements are moral or immoral. I can only speak in terms of procedure, of whether certain agreement contents necessarily conflict with the process of agreement itself. Moral advice will have normative force if and only if it provides the best semantic representation of evolutionarily fit strategies under diverse social interactions. Being clear on who counts as a concerned party to those interactions is a key part of that synthesis.

7.1.3 Homosexuality and Evolution

Earlier, I boldly asserted that homosexuality is morally permissible. Apart from the worry about pretheoretical notions, is an appeal to the morality of homosexuality consistent with an evolutionary account of morality? Any unfit allele is certainly not going to find moral support by Social Darwinists.[3] But evolutionary ethicists need not (nor should they) align themselves with Social Darwinism. Under Social Darwinism, we ought to scrap welfare programs and health programs because those who are poor or sickly are hardly fit. To have the fit take care of the unfit is to artificially bolster the continuance of unfit alleles. On the same vein, homosexuality, should it be found to have no evolutionary advantage, would be deemed immoral. As discussed in 5.3.1, the problem with Social Darwinists is not that they make bad normative conclusions from their science: they do not get their science right. The concept of progress in evolution must not be confused with intentionality. The norm in evolution is extinction, not a continuous direction to perfection. Rare alleles become common; common alleles become extinct. There is no direction or purpose in evolution.

Defending homosexuality on evolutionary grounds is thought to require evidence showing homosexuality has an evolutionary niche. Perhaps homosexuality helps keep Homo sapiens below the carrying capacity of our harsh environments, as is supposed in some bull frog populations. Such manoeuvres presuppose that group selection and Lamarckianism are correct. It accepts the view that for certain behaviours to be moral, they must benefit society in some way, and there is an intention in nature toward such moral advancement, and that humans have a moral duty to further that machinery. Whereas a morality understood in terms of individual rights eschew such talk. We do not care whether a behaviour contributes to society, let alone contributes to the individual's survival, to be deemed morally permissible: all we care about is whether or not the behaviour encroaches on the liberty of others to pursue their behaviour. Tiddly-wink playing, for example, need not be defended as having evolutionary fit to be accorded moral status. Nor do we think those who are sterile, either naturally or artificially, are immoral, though sterility has no reproductive advantage.

Where does this leave us in terms of linking morality to evolution? By showing the strategy that accords conditional cooperation is an evolutionarily stable strategy, and that an implication of this strategy is that we tolerate those who make agreements that affect no one else. This leads us back to our question: Is it the case that a society that tolerates homosexuality affects no one else – namely those who are against homosexuality? One might suppose that a society that deems homosexuality morally permissible is not tolerant of religious groups who deem homosexuality immoral. A society's being intolerant of religious groups impinges on religious freedom. Religious freedom is important since it is a subset of the important freedom to believe whatever one wants. Having the freedom to believe whatever one wants is not to be interpreted as having society endorse one's beliefs, however. Likewise, religious freedom is not to be interpreted as having society endorse one's religious views. A society tolerating homosexuality does not thereby encroach on religious freedom any more than a society intolerant of racism encroaches on the freedom of racists. It might encroach on one's desire that society abide by one's own beliefs, but such a desire was never guaranteed under religious freedom. Religious freedom allows you to believe what you want, not necessarily to have your beliefs enacted.

The problem of homosexuality for evolutionists is to describe how the behaviour persists when homosexuality is counter-productive in terms of reproductive success. This problem is irrelevant to evolutionary ethics, however. Morality as an evolutionary stable strategy must be understood in terms of algorithms, not substantive moral positions. The algorithm that has evolutionary fit, according to computer simulations, is a conditional strategy that does well through correlation. Agents abiding by this procedural strategy will have nothing to say about the immorality of homosexual activity. For the immorality of homosexuality to follow from my account of evolutionary ethics, there would have to be something about homosexuality to make homosexuals necessarily unconditional defectors. No such connection can be found.

7.1.4 Moral Standing

A concerned party is a relative notion, by the way. It is not to be equated with having moral standing. We presume we all have equal moral standing, but this sense is trivial. Someone's having moral standing does not by itself qualify her for being a concerned party. Allowing that all persons have moral status does not mean that all persons may legitimately veto my actions. A further limiting condition is needed. Of those persons who have moral standing, only those who count as concerned parties may legitimately intercede in my actions. But what is the criterion for being a concerned party?

Rather than worrying about distinguishing concerned parties from unconcerned parties, one might assert that all we need do is distinguish legitimate from illegitimate complaints, and be damned who utters them. If so, it is not a definition of a concerned party we are after, but merely a distinction between reasonable and unreasonable objections that moral agents may make. For example, take three short-listed job applicants for one job. Two will go away without the job, and one will get it. Let us further suppose that the two who fail to get the job will be adversely affected. Can we say that the two rejects are not "concerned" parties? Can we say they are not adversely and directly affected? Certainly not. Yet, all else being equal, their complaint about not being hired can have no weight, just as an Olympic competitor cannot cry foul merely for failing to win, everything else being equal. The explanation is not that we remove them from the select group of concerned parties, but that, although they are a concerned party, they have entered the negotiations under an agreement of the procedures knowing what risk that entails. Their complaint, so long as it is not directed at abuse of the procedures agreed upon, is discounted because they have already agreed to precisely those procedures. Being a concerned party does not mean one can expect anything one demands.

As put, then, we can be happy counting everyone as a concerned party, and demand instead that the claims of concerned parties be reasonable or legitimate. We could go further and stipulate that "legitimate" complaints are ones where Paretian advantage is expected. To expect to reap what is good for you, you cannot propose a bargain that is bad for the other; otherwise you will not secure agreement. If I do not want to be shot, I cannot expect you to agree to a non-shooting policy unless you too do not want to be shot more than you might want to shoot. If I do not want anyone engaging in homosexual activity, we can say I want to participate in a community where neighbours have a right to intercede in my sexual affairs. Strictly speaking this is false. What the homophobe most wants is an asymmetry, but as the prisoner's dilemma aptly portrays, she would be unwise to expect it in strategic interactions.[4] Perhaps I would be willing to live in such a community; but my wanting to interfere in your actions but not the reverse is to forgo mutual advantage: hence not "legitimate." If so, it hardly matters who is making a claim against my action, the only matter is whether the claim is legitimate.

This line of thinking gets things backwards. If you wish to borrow my lawnmower, it is understood that I have the right to decline. My reasons for declining ought to be irrelevant. I may fear alien invasions should I part with my lawnmower. You might try to talk me out of my wild fantasies, but my right to decline, *for whatever reason*, supercedes all. Likewise, we have the right to decline a gift. To suggest this right is conditional upon good reason is to deny that it is a right. In these senses of my being a concerned party, I do not require a reason to withhold my consent – legitimate or illegitimate. Similarly, we have the right to decline therapeutic treatment. The battle over whether my reasons for declining are legitimate or not ought to rest solely on whether I am deemed competent, but we cannot deem refusal itself as sufficient evidence of incompetence – as perhaps too often occurs in practice.

In other words, it is not legitimacy of the argument that matters, but the legitimacy of the person making the argument that matters – at least for those honouring the principle of consent. Recognizing this demands that we formulate a criterion to distinguish who legitimately counts as a concerned party and who does not; whose complaint counts, and whose does not. We need something much narrower than moral agency.

7.2. False Starts

7.2.1 The Harm Principle

An obvious candidate to settle who counts as a concerned party concerns the harm principle. Normally it is only those who are, or who may be, harmed by my actions who may legitimately intercede. In this sense, *A* has a duty of care toward *anyone* who is likely to be harmed as a consequence of *A*'s conduct. Feinberg stipulates two components to the harm principle: "(1) it must lead to some kind of adverse effect, or create the danger of such an effect, on its victim's *interests*; and (2) it must be inflicted wrongfully in violation of the victim's *rights*."[5] The second component is out of order, however. Presumably we understand the person's rights by checking to see whether or not she was in fact a concerned party. If she is not a concerned party to my transaction, whether her interests are adversely affected is irrelevant.

Concerning whether one's interests are adversely affected, Feinberg provides the following: "*A* harms *B* only if his wrongful act leaves *B* worse off than he would be otherwise in the normal course of events insofar as they were reasonably foreseeable in the circumstances."[6] We do not want to say that *A* harms *B* only so long as *B* is worse off after *A*'s actions, since there may be no direct causal connection between *A*'s action and *B*'s status. Nor do we want to say that *A* caused harm to *B* only if *B* would have been better off had *A* not acted, since we can imagine cases where despite *A's* harming *B, B* is in fact better off than had *A* not harmed *B*. Feinberg gives a number of such cases. For example, my mugging you may have caused you to miss a plane which subsequently crashed killing all aboard. Feinberg's "normal course of events" insertion avoids these worries. Normally, being mugged will not benefit you.

The simplicity of the harm criterion is attractive. Any uncontentious example of an immoral act will have a party designated as "concerned" in this sense and yet who did not provide proper consent. A rape victim, a murder victim, an assault victim, a victim of arson, and a victim of theft are all adversely and directly affected by the acts (and presumably did not consent). The harm definition is not sufficient for our task, however. The homophobe will count as a concerned party by the harm standard. Feinberg, I take it, tries to avoid this result by the insertion of the term "wrongful" in his definition. But whether homosexuality is wrong is precisely what is at stake. If homosexuality is wrong, then the homophobe's interference is not itself wrong. And whether the homophobe's interference is wrong depends on whether or not she counts as a concerned party.

In his definition of harm, Feinberg also includes the *intention* of the harmer. *A* may be harmed by the actions of *B*, but not in the moral sense if *B* neither intended to harm *A* nor was deemed negligent. Perhaps *A* has an interest in never hearing the word "booger." Only in the case where *B* knows of *A*'s interests, and utters the profane word anyway, can we say that *A* has harmed *B*. This helps explain why sexist and racist jokes are taboo. The common defence is that the "joker" did not intend to harm, but only to amuse. Such a defence carries infinitely less weight when the "joker" has already been sufficiently informed by the "victim" of her displeasure. And when it can be maintained that a "reasonable person" *would* be so harmed by the sexist or racist joke, uttering it anyway ought to count as an intended harm.

True, we may imagine cases where the homosexuals in question are unaware of the homophobe's attitudes, in which case their normal course of events would not – *in their estimations* – make the homophobe worse off. In a case where the homosexuals are fully aware of their busybody neighbour's homophobic attitudes, this is less likely. As with the argument against sexist and racist jokes, the homophobe can avail herself of the same defence. Clearly the homophobe's *interests* are adversely affected by homosexual activity. We can imagine her saying, "Surely those homosexuals should know that their sexual orientation is offensive to my religious values, and so their continued sexual exploits are therefore a wrongdoing." This defence is consistent with Feinberg's harm criterion and so the homophobe would count as a concerned party. The neighbour's homosexual activity could not progress, therefore, without the permission of the homophobe – permission she is unwilling to provide.

We need a criterion to show why sexist and racist jokes are an assault, but not homosexual activity. The homophobe's values are clearly infringed, and this adversely affects her. If the concept of harm is linked to one's values this closely, then our moral structure collapses into subjective relativism. I cannot interfere with your actions unless I deem it ok for me to do so. If there is no restriction on what I deem to be "ok" for me to do, such a guide is useless.

To argue that the homophobe is not in fact harmed, by the way, is to have a narrow definition of harm. It is no use arguing that the homophobe is not *directly* affected, for it is precisely the homosexual sex that affects her, and not some fallout from that act. Granted, we might suggest the effect on the busybody is not direct in the sense that the harm she endures is not actually a result of her neighbour's activities. She is bothered by homosexuality in general, and her neighbours' activity is merely an instance of that. But such manoeuvres are pedantic, at best. That I am generally loath to being raped does not mean I am not directly affected by a rapist's raping me, or a rape happening near me.

We might say that an assault victim is harmed irrespective of her attitudes toward assault. Perhaps this is not true for all. If I have a favourable attitude to my being assaulted, it is not clear that assaulting me would harm me. Still, a distinction can be made between cases where one's attitude is irrelevant to the being harmed and cases where one's being harmed is dependent on one's attitude. The homophobe's harm is of the latter sort. The harm she endures is solely connected to her negative attitude toward homosexuality. This distinction may lead some to designate "harm" only in cases where attitude is irrelevant. By defining a concerned party in relation to this narrower understanding of harm, we would effectively rule out homophobes as concerned parties to the affairs of homosexuals. It would do so at a cost, however. Being harmed by sexists and racist jokes, or poisoned environments in general, would not qualify one for being a concerned party. Surely members of minority groups so targeted ought to count as concerned parties.

Similarly, some might suggest that harm must be understood in physical terms only, but not counting as harm psychological harm, financial harm, or libel, is absurd. We might wish to deny the legitimacy of the homophobe's complaint, but to deny her interests are affected is implausible. The harm principle alone is not sufficient.

7.2.2 The Wellsian Position

Some may contest that the worry about defining a concerned party is idle, not because we are all moral agents, but because our social dependence is such that we are each invariably affected by others' actions. Since H.G. Wells, it is customary to warn time-travellers not to touch anything, for even a single footprint can alter the course of history.[7] Holistic environmentalists urge the same warning in the present: a butterfly's flutter in Brazil can cause tornadoes in Taiwan, or so the story goes (although I have not heard meteorologists explain weather patterns in this way). Similarly it is argued that we are not atomistic individuals, that no action you can do can leave others unaffected.[8] If so, defining a "concerned party" as someone who is affected by one's action is idle since we all are, inevitably.

Let us grant that we are all concerned parties in this large sense. Ethical principles purport to be universal principles that apply to all people (minimally) in virtue of their being people (or minimally in virtue of their having features in common with all people, which may include some non-persons). That is, no one is to be excluded. This is true in regards to who counts as a moral agent, but, as mentioned above, moral agency is a broader concept than being a concerned party. To confuse the two would be to demand the impossible task – at least in terms of the principle of consent – of eliciting consent from everyone in the world prior to any action. Contractarian ethical theorists purport to manage such universalism by an appeal to what hypothetical agents in ideal situations would agree to. Such bargaining concerns principles and procedures, not particular actions. What if the procedure these ideal agents agree to under such circumstances is something akin to our applied principle of consent? Then we are no further ahead. Is it the consent of *everyone*, or of only some much smaller group of designated persons? The decision on how to define a concerned party needs to be resolved even at the *ex ante* bargain.

That we may all be affected by any given act, even if true, is idle, in any event. Meaning has been stretched to implausible lengths. Even if the entire world were an illusion, we still want to distinguish between what we are wont to call mirages and what we are wont to call reality.[9] Likewise, even if we are all affected by each action, the rape victim is affected more by the rape than by someone having coffee on the other side of the country. Even if we are all "concerned parties" to everyone's actions in the grand Wellsian sense, we still want to make a distinction between the sort of concerned parties that matter from those that do not. Let us call only the latter "concerned parties." (The former we shall name "moral agents.") When we speak of a concerned party we are limiting ourselves to those directly involved, and not to those affected merely in the Wellsian sense.

Coupled with the Wellsian picture, some may understand consent as the default. If so, we may presume everyone consents to all of our acts *unless* someone pipes up and complains. That no one complains that I drink coffee does not mean no one is a concerned party. Simply, they have nothing (yet) to complain about. When assaulted, on the other hand, we have something to complain about. The danger in adopting a default approach to defining a concerned party is twofold. First, someone's piping up may be inhibited by an effective threat or a feeling of powerlessness. Lacking a voice may be a symptom of abuse, not an indicator of the absence of abuse. That is partly the trouble with our difficulty in prosecuting date rape and lowering instances of spousal and child abuse.[10] It is worse if even the piping up is discounted by a presumed overriding consent. We cannot tolerate the *presumption* of consent to sex, and an effective criterion of a concerned party ought to rule this out. Secondly, we open ourselves up to being curtailed by the merest whim. That I do not like foreheads pierced would be sufficient to stymie a recent trend. That the homophobe complains would be sufficient justification to prevent homosexual activity.

In sum, we cannot tolerate everyone's being a concerned party in the Wellsian sense. It is not simply a matter that anything is approved unless someone objects. Certain people's objections may be irrelevant. What we want is a criterion to make that distinction.

7.2.3 The Use Definition

We might consider a more Kantian definition of a concerned party: *Someone is a concerned party only so long as the affecter cannot fulfil her desires without using the affectee's person or property.* By this definition, we move away from the effects on others and examine, instead, the nature of the act itself. An advertiser may use the image of a famous athlete in his commercial. The athlete is a concerned party in this case. Therefore, the advertiser requires the consent of the athlete. My use of your lawnmower entails that you are a concerned party by this definition.

Notice that persons may be used without suffering adverse effects, and in some cases persons may be benefited despite their not consenting to such use. An example of the first case is when *A* borrows *B*'s lawnmower without *B* knowing, and refills the gas, tops the oil, and sharpens the blades before returning it. An example of the second case is when *A* mugs *B*, and in the process causes *B* to miss her flight that subsequently crashes. Despite *A*'s intent, *B* is made better off by *A* than she would have been absent *A*'s intervention. If we focused merely on whether one was made worse off, in neither case could we complain about *A*'s action. Under the Kantian use criterion, however, so long as *A* acts without *B*'s consent, *A*'s act is immoral.

Notice that the use criterion avoids the need of Feinberg's "normally considered" clause. Also, the use definition will show us that the homophobe is not a concerned party in the homosexuals' affairs. Sex between two people other than the homophobe neighbour does not require the use of the homophobe neighbour. Persons may engage in homosexual sex with or without their neighbour being a homophobe. Therefore the homophobe does not count as a concerned party – despite her suffering adverse effects. Moreover, all the standard immoral acts are cases where someone was used without his consent. Murder, theft, assault, rape, and arson are all instances that use the targeted victim; therefore, the targeted victim has a right to be consulted about the proposed act. Should the act progress without that consent, the action is immoral.

Although promising, the use definition will not do. If Smith decides to shoot off his gun in various directions, the use of Jones, Smith's proximate neighbour, is not needed for this pursuit.[11] Whereas the rapist cannot proceed without the rape victim, Smith can proceed to shoot his gun if Jones is not present. Therefore Jones would not count as a concerned party by this definition, despite Jones's chances of being shot by Smith's activity – an unequivocal example of suffering an adverse effect. If Smith were using Jones as target practice, that would be a different matter. In the case under review, Smith is simply oblivious of Jones. Jones need not be there for Smith's purposes. Clearly, we want a definition of a concerned party that would say Jones is a concerned party in this case. The use definition cannot do so.

7.2.4 Conjunctive and Disjunctive Definitions

The harm definition allows busybody interlopers to intercede in otherwise perfectly moral behaviour. The use definition precludes otherwise perfectly sensible complaints of concerned parties. Can we simply conjoin the two? Neither a disjunct nor a conjunct of the two are satisfactory. In the disjunctive version, any person will count as a concerned party if any one of the conditions is met. This would help the neighbours of the random shooter, for they would now become concerned parties due to the harm condition of the disjunct, but it could not rule out busybodies, for their interests, too, are adversely affected. In the conjunct version, although one horn of the definition applies in both cases, in neither case do both apply. The result is that although busybodies no longer count as concerned parties, nor would neighbours of random shooters.

The two conditions cannot be happily joined into one overarching criterion. One component satisfies our understanding in the one case, and the other component satisfies our understanding in the other, but there is no overarching guide to tell us which condition properly applies in which case: precisely our goal.

7.2.5 The Aiming Condition

Under the use condition, victims of random shootings would not be concerned parties. This is clearly inadequate. We need to distinguish the adverse effect of the homophobic busybody from the adverse effect of the victim of random shooting. In the case of Smith and Jones, it is clearly the bullet fired from the gun that strikes the neighbour, or potentially strikes the neighbour, that is the problem. It is not merely the neighbour's firing of the gun. Firing a gun under controlled conditions in a firing range would not require the neighbour's consent. Compare this circumstance with the homophobe's complaint. She is, euphemistically speaking, arguing against the firing of the homosexuals' guns in any case – even in the privacy of their own home. We could sympathize with the homophobe neighbour if sex was occurring in her yard. This would be analogous to firing a gun *into* her yard, or, for that matter, drinking espresso in her yard when she wants no one doing anything in her yard without her consent.

We would say, then, should we desire a general rule, that firing guns by civilians intentionally or not is fine if and only if it is not aimed at any person or the exclusive property of any person without prior consent. This makes sense, at least. It does *not* make sense to apply this sort of general rule to homosexual activity since it is not *aimed* at anyone in the first place – even unintentionally. Other than masturbation (and perhaps even then), sex is obviously aimed at someone. Morally permissible sex does so with the consent of all those towards whom the sex is aimed. Sex aimed at those who do not consent is assault. For those towards whom the sex is not aimed, their consent is not needed. Assuming this is understood, "sex" then captures the activity of the group involved. Their "aim" is internal, not external. That is, the group does not aim at anything beyond the group itself. Admittedly, this is a very cumbersome manner of speaking and that is partly the point.

To aim at something is usually understood as an intentional act. Smith might protest that he did not aim at Jones, for example. So, to avoid the oxymoron of an unintentional aiming, we have to here understand "aiming" more loosely. Granted, the random shooter is not *intentionally* aiming the gun at Jones, but so long as the gun is aimed at Jones, or Jones's property, Jones is a concerned party. Admittedly, Jones may have kin who are not properly to be regarded as Jones's property, and some might complain, then, that Jones's kin are not concerned parties. Perhaps Jones's wife and daughter are playing bocce in the yard. We do not want to be forced to call them "property of Jones" in order to protect them from stray bullets. So let us read "Jones" as simply a place holder. Any person being shot at can be represented by "Jones." Thus the concern is *a* person and that person's belongings, and not a specific person and that person's belongings. The aiming condition, therefore, will take the following form: *Someone is a concerned party only so long as the affecter aims, intentionally or otherwise, at that person or that person's property.*

Will this definition pass our test cases? Yes. Homosexuality is not aimed at the homophobe. The homophobe therefore would not count as a concerned party. This is as it should be. The neighbour of the random shooter, however, would count as a concerned party. This is also as it should be. The gun is (at times) aimed at him. Moreover, standard immoral acts such as murder, assault, theft, rape, and arson all affect concerned parties who have a right to veto those actions aimed at them.

Although the aims definition satisfies our test cases, it is less satisfactory in other cases. Second-hand cigarette smoke is not aimed at anyone in the same sense as the bullet of a gun is. Therefore non-smokers would not count as concerned parties to smokers' actions by the aims definition. Surely they ought to so count. To modify the aims criterion to include the snaky trajectories of second-hand cigarette smoke would have to focus on whether or not another party is in fact affected by one's action. This modification collapses the aims condition into the harm principle. The homophobic neighbour returns as a concerned party.[12]

7.2.6 The Two Adverse Effects Rule

There are two types of adverse effects: physical and psychological. Physical harms include harms to the body and one's property (including financial harms). Psychological adverse effects include any other felt harms which may include social harms, such as libel,[13] and "spiritual" harms where the actions of others are deemed offensive to one's religious convictions, for example. The advantage of the two-effects rule is to note that the criterion for a concerned party will differ according to whether the adverse effect is physical or psychological. Anyone suffering or expected to suffer physical adverse effects is a concerned party and her consent is required. The neighbour of the random shooter is clearly a concerned party in this sense. If the adverse effect is psychological, then *B* is a concerned party to *A*'s action only so long as *A* cannot fulfil her desires without the use of *B*'s person or property. The homophobic neighbour fails to be a concerned party on the first condition, but she also fails on the second. Therefore, she would not count as a concerned party; her lack of consent is irrelevant. The person suffering second-hand smoke is a concerned party, since carcinogens are physically adverse. Likewise, Jones is a concerned party to Smith's random shooting, given the physical adverse effects Smith is at risk of suffering. The two adverse effects definition of a concerned party, then, avails itself of both the harm condition and the use condition, without collapsing into a simple conjunction or disjunction. It takes the following form: *Anyone suffering or expected to suffer physical adverse effects is a concerned party. If the adverse effect is psychological, then one is a concerned party only so long as the affecter cannot fulfil her desires without using the affectee's person or property.*

We are getting closer, but this will not yet do. Stompin' Tom Connors was inducted into the Canadian Country Music Hall of Fame against his will. Presumably such an honour had no adverse effects, either physical or psychological, yet few would deny he had a right to decline the offer. Such a right presupposes he counts as a concerned party. After all, I did not have the right to veto it, but Stompin' Tom certainly did. Therefore a concerned party need not be restricted to only those who are adversely affected, physically or psychologically.

One might argue that Tom was psychologically upset by the honour, otherwise he would not have declined. The danger in conceding this is that we would permit anyone's declining as evidence of a psychological adverse effect. Since we are seeking to distinguish when objecting to a negotiation is legitimate, this move will hardly do: it amounts to admitting that an objection is legitimate so long as one objects – hardly satisfactory. Another problem lurks by doing so. Can someone be adversely affected without minding? Can we not envision cases where justice requires compensation to a clueless victim? My concern at present, however, is to discover an effective way of ruling out a self-proclaimed concerned party, or at least to distinguish legitimate from illegitimate self-proclaimed concerned parties.

7.3. The Solution

If we concede that Stompin' Tom is psychologically adversely affected, the use condition can still apply to make him a concerned party. His use broadly conceived includes his mere existence: something necessary for the Canadian Country Music Hall of Fame's intent to induct him. This would follow, however, only so long as Stompin' Tom is deemed to be psychologically affected. Whereas we should now recognize that the use condition should trump the psychological condition, and not be dependent upon it. That is, we should replace the psychological adverse effect condition with the use condition, while keeping the physical harm condition. We get the following and final definition of a concerned party:

Anyone suffering or expected to suffer physical adverse effects is a concerned party. Otherwise, one is a concerned party only so long as the affecter cannot fulfil her desires without use of the affectee's person or property.

Let us call this the *Harm-Use* definition of a concerned party. By the Harm-Use criterion, Stompin' Tom, regardless of whether he is psychologically affected, is a concerned party to the Canadian Country Music Hall of

Fame's action. So long as second-hand smoke causes physical damage, non-smokers are concerned parties to nearby smokers' actions. The homophobe busybody is not a concerned party by the Harm-Use definition, but the victim of the random shooting is. Moreover, victims of all the standard immoral acts, such as murder, theft, arson, rape, and assault, count as concerned parties by this definition.

A caveat is needed. Consider racist and sexist jokes and other cases of poisoned environments. In this case, let us imagine *A*'s joke is not aimed at *B*, a member of a recognizable minority, but at the general group to which *B* belongs. *B*, therefore, need not be present or even exist, for *A*'s joke. Therefore it would appear *B* is not being used in the requisite sense. Since it is also the case that *B* is not physically harmed by the joke, it would appear that the Harm-Use criterion could not count *B* as a concerned party. If *B* is not a concerned party, her complaint about the racist or sexist joke would be deemed morally irrelevant.[14] Since this will not do, we need to expand the meaning of "use" to include any member of a group that is used. In this case, the sexist/racist joke cannot proceed without the targeted group, and since *B* is a member of that group, *B* will count as a proxy representative of that group. The point here is that *B* may rightly complain, not that *B must* complain. If *B* is not bothered by the racist/sexist joke, *A*'s action will be deemed inoffensive.

It is not impossible to find further difficulties. Consider Malcolm Lowry's tale of the ex consul who lets his garden go untended.[15] This bothers his neighbours who leave him a threatening sign. Their yards are trim, their houses kept. Can the neighbours rightly complain according to our Harm-Use definition of a concerned party? This case is more ambiguous. So long as the ex-consul's letting his house and yard rot does not cause physical harm (which it does not, at least narrowly defined), nor does it require the use of his neighbours (which it does not), their complaints cannot count as the voice of concerned parties. We cannot define concerned parties according to entitlement.[16] If we define "entitlement" synonymously with "legitimacy," this simply collapses into a tautology. The only legitimate complainants are concerned parties and concerned parties are those whose complaints are legitimate. Not very helpful. For the liberal minded, the link to an entitlement theory is all very well. For the communitarian minded, it is not. Property may be partly controlled by the interests of the community in which we live, and if we define concerned parties by our final definition, we will preclude communitarian arguments at the outset.

On the other hand, so long as our definition of "physical" allows the prospect of financial harm, the neighbours' protestations may be justified even on the liberal reading. We need not suppose the complaints of the ex consul's neighbours are merely aesthetic. Aesthetics cost. The ex consul's sloth may lower the resale value of the neighbouring homes. Also, the ex consul's garden's going to seed can contaminate their gardens. That would be analogous to second-hand cigarette smoke, and their complaint is justified.

Notwithstanding these complications, I submit my Harm-Use definition of a concerned party to the public:

> *Anyone suffering or expected to suffer physical adverse effects is a concerned party. Otherwise, one is a concerned party only so long as the affecter cannot fulfil her desires without use of the affectee's person or property.*

7.4. The Role of *Ex Ante* Consent

The Harm-Use definition of a concerned party should not be thought to do all the needed moral work. It must be understood within the larger framework of the principle of consent: *Any act is morally permissible if and only if all competent, suitably informed, concerned parties voluntarily consent.* But understanding a concerned party within this larger framework creates other worries. As put, the moral condition of agreement among all concerned parties (1) seems to preclude many transactions we are inclined to label morally permissible, and (2) seems to permit actions we are disposed to call immoral.

1. Consider the following example of the first worry. An association, church, or club ejects one of its members for policy violations. In such a case, the following conditions seem to be met: (i) the act is deemed morally permissible; (ii) the ejected member is a concerned party by the Harm-Use definition; and (iii) the ejected member would not (we can imagine) agree to being ejected if she had any say in the matter. But the principle of consent would indicate that since the excommunicated member is a concerned party, yet did not consent to such treatment, the action ought to be deemed immoral. So either the large claim that morality entails agreement among all concerned parties is false, or one of these conditions is not met.[17]

The answer is that condition (iii) is not met, or is at least misleading. The concept of consent is ambiguous. When I lose a bet and have to fork over my money, I do not agree to do so at the time. In fact, my not agreeing to hand over my money at that stage can hardly count as absolving moral responsibility. When we speak of consent, therefore, we have to distinguish between *ex post* and *ex ante* agreement.[18] When I make the bet, I agree *ex ante* to pay up if I lose despite my likely *ex post* disinclination to do so at the time. Without that *ex ante* consent, no one would negotiate with me. Similarly, those who enter clubs, churches, and associations take on *ex ante* agreements to abide by the rules of the club, church, or association, and to accept the consequences for violating them. That they do not agree with the consequences *ex post* is irrelevant, even though they are concerned parties. (If they do not agree with them *ex ante*, they will not be motivated to join in the first place.) My concern in the preceding chapter was not to call *ex ante* agreements non-binding, but to solve the incoherence problem by understanding the normative desirability of *ex ante* agreements in terms of evolutionary fit.

There are other types of cases. (a) Consent theory would seem to outlaw punishment by the state, since no one consents to be born and raised in a state. The same holds for self-defence. We normally believe it is permissible to kill those who are about to kill innocents. Killing unprovoked aggressors, however, would be done without the aggressor's consent, and so consent theory would seem to frown upon self-defence.[19] (b) For another kind of case, consider the concept of fair play. Consent theory cannot seem to account for the intuition that certain obligations are thrust upon us without our consent. Consider when someone does us a favour. Like it or not, civil society foists upon us the obligation to repay the favour in kind, or at the very least, to express gratitude. Take the case where *A* shovels the snow of a driveway that enables *A* and *B* to get to town to get food. We assume *B* is obligated to do the digging at the next snowfall, if he can. At the very least we expect *B* to thank *A*.[20]

Neither kind of example stymies consent theory. Consent theory's merit is conditional upon the compliance of others. Absent that, all bets are off. In case (a) punishment or retaliation is deserved only in the case where the "aggressor" has already violated (in the case of punishment) or is presently in the act of violating (in the case of self-defence) the principle of consent. Violating the principle of consent entails resigning from the protection the principle of consent normally affords one. Morality, recall, is conditional, not unconditional. There are cases where the conditions fail to apply. The correct response to a defector is defection. Consent theory is not forbidden from saying that.

Case (b) concerns the problem of free riders. There is a benefit to *A* irrespective of *B*'s presence. That is what motivates *A* to do the shovelling. *A* might rather not benefit *B* if *B* remains a free rider. Likewise, *B* would either have to do the shovelling himself or go without food until the snow melts. Without *A*'s presence, we would not say *B* is obligated to shovel. So how does an obligation to *B* occur when, without *A*, there would be no obligation, and without *B, A* would be doing exactly the same chore? Consent theory does not mandate that certain kinds of agreements must be made: only that once you make them, you should honour them. But does the fact that *A* and *B* share the same driveway entail an implicit agreement concerning driveway use between *A* and *B*? One would certainly assume so, but to rely on implicit agreements in hard cases would cheapen consent theory. No *ex ante* agreement can be so detailed as to include snow shovelling arrangements. If *B* refuses to admit any such implicit agreement, there is no empirical matter to which *A* can point to change *B*'s mind. All that one can say concerning the convention of fair play is that those who adapt to certain correlated conventions tend to do better in the long haul than those who do not. But nothing in this rules out niches to break the convention. Consent theory focuses on what sort of interpersonal structural arrangements tend to do well for individual agents in the long run. Consent theorists call those structures the basis of what we mean by morality. Agents who fall into this pattern may well feel an "obligation." Such a phenomenon is not part of consent theory. It is an epiphenomenon. Consent theory remains irrealist. There is no "obligation" in the way that moral realists mean. There is a pull, and we do well, generally speaking, to be so pulled, but concerning looking for an extra ingredient – the nature of the obligation independent of the correlated convention – consent theory remains skeptical.

2. A more common worry follows from imagining cases where someone is not deemed to be a concerned party by the Harm-Use criterion yet toward whom we are morally obligated. Consider the case of helping someone from drowning. If Smith, who is drowning, calls out to Jones for help, and no one else is around, and Jones can save Smith without harm to Jones, and Jones can reasonably predict that no harm to Jones will ensue from Jones's rescuing Smith, we feel it would be morally reprehensible if Jones nevertheless declines to help, or offers to help only for a fee. But Smith would not count as a concerned party to Jones's actions in this case if we rely on the Harm-Use definition. Jones can carry on with his affairs without Smith, and Smith is not made worse off by Jones. Smith is drowning with or without Jones's presence. Thus Smith would not be accorded any moral obligation toward Jones by the principle of consent. So either the principle of consent is wrong, or there is no moral obligation to help, or Smith really is a concerned party to Jones's action in which case the Harm-Use criterion of a concerned party is flawed.

The larger answer to this problem I reserve for the following chapter. The *rational* answer is familiar. Gauthier, for example, points to *ex ante* consent. Once we retreat to the prospect of hypothetical *ex ante* agreements binding our *ex post* actions, we could retreat further to an ultimate hypothetical *ex ante* agreement concerning what sort of moral rules we would be willing to be bound by, and see if saving persons under the situation described (at least) would be the sort of thing to which reasonable self-interested agents would agree. And this works for those who would agree. But they are not the ones who worry us.

Of course, relying merely on naturally occurring (rather than socially inculcated) good will, will not suffice. If duty is predicated on the existence of the requisite good will, the absence of good will absolves one of a duty. Therefore, we would not be able to criticize Jones for failing to help if Jones does not have the requisite good will. But it is false to suppose that merely having the requisite good will to help others is sufficient to help others. Other desires may compete. So we need an argument to show why persons who have some interest in helping others may be motivated *ex ante* to consent to being morally bound to help under specific circumstances.

An assurance argument can do the job. This is not to be confused with an insurance argument. An insurance argument says you would be willing to be bound to help others so long as others are also bound to help you. Persons *might* agree to be bound by a duty to help under limiting conditions for this reason, but the benefits of being helped on the few occasions where I would need help must outweigh the cumulation of small costs from all the times I would need to help others. Since the probability of my needing help will be greatly outweighed by the probability that someone else will need help, it is not clear that the final value is positive. If, contra Singer,[21] the limiting conditions specify a proximity rule, the insurance argument will carry more weight. That is, the number of times I will be required to help will be greatly diminished if I must help only those others in close proximity to me, not those farther removed however dire their conditions may be. Since my needing help will still be met by those in close proximity to me (who have also bought into the insurance scheme), the cumulative costs of helping others decreases significantly, while the benefits of being helped remain the same. Such an adjustment may be sufficient to make my buying into a helping scheme cost-beneficial. Perhaps such considerations would move purely rational agents into acquiescing to a system where helping others is mandatory, but few believe such calculations really explain our penchant to help. Angry gods could indeed explain thunder, but merely finding a consistent explanation does not mean one has found the right explanation. An assurance argument, on the other hand, need not stray so far from our natural ascriptions.

The assurance argument goes as follows: Despite our sharing a preference for something, we may require an extra motivation to act on it. It is conceivable that despite our wanting to do *x*, we know that left to our own devices we probably will not do *x*. Wanting to exercise is not a sufficient condition for exercising, for example. This is the problem that an *ex ante* agreement attempts to solve. Notwithstanding our desire to make agreements, we also know that we will be tempted by the lure of profit or sloth to unilaterally renege. Likewise, despite our natural inclinations to help, we may require assurance that our help will (a) not be wasted, and (b) not disadvantage us. Both assurances may be satisfied if we *ex ante* enter a pact with other like-minded individuals similarly thwarted from helping when they desire. If we internally hard-wire ourselves to help, this will certainly increase the odds of successful helping, nor will it disadvantage us, since we are all (or most of us) equally disadvantaged by our generosity.[22]

What has this achieved? (1) It admits that we have natural (albeit limited) sentiments to help. (2) It shows how we may nevertheless require an *ex ante* agreement to commit ourselves to help despite our natural sentiments. (3) The duty comes from being bound to an agreement, and so we have no cause to worry about question-begging or naturalistic fallacies. And (4) the agreement is rationally motivated for all persons who satisfy the two conditions of (a) having natural sympathies to help, and (b) are thwarted from helping voluntarily due to the assurance problem.

Such reasoning does not extend to forcing people to abide by these rules who would not voluntarily agree to them in the *ex ante* position. Refusal is possible so long as there is no natural inclination to help, or that whatever natural inclination exists does not outweigh competing desires. It is sufficient for now, however, to show that it is not paradoxical to consent to rules that would force us to comply with these rules even when, at the time of being forced to comply, we would rather not. Chapter 8 will offer an evolutionary account of helping behaviour that does not rely on rational consent.

A variety of other kinds of counterexamples to consent theory may be offered. For example, consent theory morally permits buying another person's vote, renting another person's sexual organs, and purchasing someone who is willing to contract into lifetime slavery (perhaps she needs the money to pay for an operation necessary to save her brother's life).[23] But – so the argument goes – these agreements are morally objectionable, whereas consent theory claims that no agreement in itself is morally objectionable. If we think so, it is because the procedure of the agreement is faulty. A similar reply to the obliging stranger case (6.6) can be made with selling oneself into slavery. We cannot imagine a competent individual would consent to such a thing. But so long as this is given, and no one else is a concerned party (no dependents, for example), then the transaction is morally permissible.[24] That this seems counterintuitive is only because we cannot imagine the case of any *competent* individual actually consenting. In the case of renting one's body parts for sexual gratification, consent theory will say society is wrong to suppose that is immoral, so long as the consent of all competent, concerned parties is voluntary, informed, and entail no externalities. Perhaps those conditions are not always met. Even their never being met would not count against consent theory. We would explain why the activity is not moral by pointing out how it fails one of consent theory's conditions. Selling one's vote can be understood as immoral, though, by noting that a concerned party is excluded in the vote transaction. Votes are, first off, positive rights of citizens provided by the governing body to each voter equally. Part of the stipulation of this positive right is a one-vote-per-person rule. Selling votes violates that rule, which violates the agreements made between voters and governments. Consent theorists, thereby, can explain why vote selling is a moral infraction as easily as they can explain why the agreement among a gang of thieves is not sufficient to make their thievery moral.

But what if some individuals did not consent, even *ex ante*, to the institution of a one-vote-per-person rule? For those who did not, or would not, so consent, all bets are off. Should consent theorists call those non-consenters immoral? As Harman notes, to complain about those who reject our *ex ante* agreements is not to make a moral judgement, since they are outside the necessary conditions where moral appraisal makes sense.[25]

7.5. Summary

This is one of those specialized chapters, answering a very specific problem. It is not a problem that has received much attention to date, because few ethical theories make such bold use of consent that I recommend. But once we define moral actions according to consent theory, we need to be very clear whose consent we are talking about. In this chapter, I hope to have clarified in what sense not everybody's consent matters. To distinguish consent that matters from consent that does not matter, I offered the following criterion:

> *Anyone suffering or expected to suffer physical adverse effects as a result of an action is a concerned party to that act. Otherwise, one is a concerned party only so long as the affecter cannot fulfil her desires without use of the affectee's person or property.*

If you satisfy that condition on a particular occasion, what you say matters. If you do not satisfy that condition on a particular occasion, what you say about that occasion does not matter. Sorry.

For those who feel consent theory leaves out too much, specifically the demand that we morally ought to want to help others, a larger argument is required. I offer that in Chapter 8.

Notes

[1] This is Kai Nielsen's injunction in his "Capitalism, Socialism, and Justice," in T. Regan and D. Van DeVeere (eds.) *And Justice for All* (Totowa, N.J.: Rowman & Allenheld, 1982), 264–86.
[2] Technically, a "homophobe" is one who fears homosexuals. One may deem homosexuals are immoral without being a homophobe, as I might deem a pedophile immoral without being afraid of a pedophile. Likewise, one may fear homosexuals without believing homosexuals are immoral, as I might fear dogs without thinking dogs or dog owners are immoral. Still, the term is a handy referent to capture the group who believe homosexuality is immoral and I mean it in this loose sense.
[3] Herbert Spencer, "Progress: Its Law and Cause," *Westminster Review* 9 (1857): 445–85. Herbert Spencer, *First Principles* (London: Williams and Norgate, 1862).
[4] For my discussion on strategic, rather than parametric choice, see 6.2.1. For a more detailed account, see David Gauthier, *Morals By Agreement* (Oxford: Oxford University Press, 1986), 60–82.
[5] Joel Feinberg, "Wrongful Life and the Counterfactual Element in Harming," in his *Freedom and Fulfillment: Philosophical Essays* (Princeton: Princeton University Press, 1992), 3–4.
[6] *Freedom and Fulfillment*, Feinberg, 11 (italics removed, typo corrected).
[7] H.G. Wells, *The Time Machine* (Dover Publications, [1895] 1995).
[8] For example, Charles Taylor, *Philosophy and the Human Sciences: Philosophical Papers 2* (Cambridge: Cambridge University Press, 1985), 187–210.
[9] This is the Natural Language School argument against philosophers' demarcations. See, for example, O.K. Bouwsma, "Descartes's Evil Genius," *The Philosophical Review* 65 (1949): 141–51.
[10] See Lois Pineau's important complaint: "Date Rape: A Feminist Analysis," *Law and Philosophy* 8 (1989): 217–43.
[11] This counter-example was offered to me by John Woods.
[12] One could say second-hand smoke violates property rights in the sense that one has ownership of one's own lungs, and second-hand smoke contaminates that use. A difficulty, though, is that ownership in this sense may also be extended to one's moral sensibilities. One owns one's personal moral outlook and should that outlook find homosexuality to be a sin, then society's sanctioning homosexuality can be seen to be as much of a contaminant to one's property rights as second-hand smoke. To ward off that extension, the harms condition needs to be resurrected.
[13] Some social harms such as libel may be financially calculable.
[14] This counter-consideration to the Harm-Use definition was raised by an anonymous referee.
[15] Malcolm Lowry, *Under the Volcano* ([Jonathan Cape, 1947] Penguin, 1980), 132.
[16] This was my belief in "How to Blackmail a Contractarian," *Public Affairs Quarterly* 13/4 (1999): 352.
[17] I wish to thank an anonymous referee for raising these worries.
[18] As David Gauthier makes explicit in *Morals by Agreement*, 9–10.
[19] I was alerted to both worries by Thaddeus Metz.
[20] See for example, George Klosko, "Political Obligations and the Natural Duties of Justice," *Philosophy and Public Affairs* 23/3 (1994): 251–70.
[21] Peter Singer, "Famine, Affluence, and Morality," *Philosophy and Public Affairs* 1 (1972): 229–43.
[22] We are not disadvantaged relative to others. We may be in non-relativistic terms.
[23] Thanks to Thaddeus Metz for raising these examples.
[24] This is different than Mill's response. He believed the preservation of liberty counts against paternalism. Since selling oneself into slavery is counter to preserving one's liberty, we have non-paternalist grounds to prevent the agreement. J.S. Mill, *On Liberty* (Hackett Publishing Co., [1859] 1978), ch. V, 101.
[25] Gilbert Harman, "Moral Relativism Defended," *Philosophical Review* 84 (1975): 3–22.

Chapter 8

SUFFERING AND INDIFFERENCE

8.1. Introduction

A common criticism of contractarianism is that it cannot accommodate duties of beneficence: it is too limited in scope.[1] Although my consent theory distinguishes itself from contractarianism in terms of offering a more naturalized epistemology, it does not avoid the worry concerning indifference to suffering. Perhaps, even, it faces the problem of indifference more severely. Contractualists do not have this difficulty, for they claim to embed requirements of aid into their theories. For example, Rawls's difference principle ensures equitable transfer of goods for the relief of suffering and this is grounded by what individuals would agree to from behind the veil of ignorance.[2] The difficulty with this attempt is that it derives morality from normatively rich beginnings. Idealized agents starting from fair bargaining positions will easily (too easily) yield fair outcomes. What is needed is to show why actual agents would feel bound to the agreements of these artificial agents. That I would agree to marry someone else in alternative circumstances is not justification to divorce my current wife. Scanlon argues that limited beneficence is part of our moral duties since it is a principle that no one who shares the goal of finding a common principle of ethics could reasonably reject.[3] The difficulty here is that having a goal of coming to an unobjectionable agreement does not by itself dictate what the content of that agreement is.[4]

A's being indifferent to *B*'s suffering (so long as *A* is not the cause of *B*'s suffering) satisfies the consent theory credo. For Joseph to elicit an action by Mary, Mary needs to consent. If Mary does not consent, Joseph cannot force the desired action from Mary. For example, if the action Joseph wants from Mary is for Mary to hand over her wallet to Joseph, and Mary does not consent, Joseph cannot morally force Mary to do so. We call that mugging. The form is immoral for consent theorists wherever it occurs. Joseph's demanding help from Mary (or society's demanding Mary help Joseph) takes precisely the same form. Merely changing "hand over the wallet" to "pull me out of this water" does not alter the pattern. Mary must consent. The fact that most of us would consent is beside the point (and at any rate would not justify why there is a moral duty to do so). All we require is one situation where Mary is utterly indifferent to the plight of Joseph, and so declines the offer. To force the indifferent Mary to help Joseph notwithstanding Mary's non-consent is a violation of the consent theory credo. Indifference is morally permitted.

Although it is unrealistic to expect a moral principle to eradicate all causes of suffering, many suspect it is not beyond a proper moral principle to try to eradicate suffering when we find it – whatever the cause. My intent in this chapter is to explain how our moral sentiment concerning helping others is consistent with evolution-based consent theory.

8.2. Agreements

8.2.1 Abiding by and Making Agreements

Although we may have a reason to make an agreement, we also have a reason to renege on that agreement. If morality is operationally defined as compliance with agreements made when there is no harm to oneself, then morality and rationality appear to part ways. To maintain the position that morality is in one's self-interest after all, consent theory highlights the evolutionary benefits of adopting a conditional cooperative disposition, a disposition to comply with the agreements we have found reason to make when interacting with others who are also prone to comply.[5] Since these agents do better than other types of agents in terms of evolutionary fitness, a gambling man's argument can be made for being moral. The success of constrained maximization requires internalizing this moral disposition, to hard-wire oneself, not specifically to avoid the temptation to defect, but to avoid having others believe you will succumb to the temptation to defect.

What does not follow, however, is that conditional cooperators (CC) will act in ways we commonly deem moral. Internalizing our moral disposition to *comply* with agreements does not in any way commit us to *make* certain agreements. All we have (which is quite a lot, really, and I do not wish to underestimate that, though it is not my concern here) is a statistically based reason for you to internalize a disposition to abide by agreements you deem in your interest to make. Despite showing that you ought to internalize a disposition to abide by your agreements made when interacting with others not likely to defect, there is no room to argue that there are certain kinds of agreements you ought to make. The argument for conditional cooperation has only limited effect.[6] If you do not happen to see any benefit to you in making and keeping the agreement, there can be no compunction for you to make it.

Common lore has it that some agreements ought to be made.[7] For example, if someone drowning asks for assistance, we presume ourselves under a moral obligation to assist if we can. Consent theory's emphasis on the moral wager to internalize dispositions to be compliant cannot share such presumptions. If someone "agrees" to help, good. If someone does not "agree," then what can be said? What moral value can consent theorists give to those who remain indifferent to suffering? None, it would appear. So long as there were no prior agreement to relieve suffering, no one will abrogate consent theory principles by remaining completely indifferent to suffering.

8.2.2 *Ex Ante* Agreements

Notice that Gauthier's distinction between *ex ante* and *ex post* agreements does not impinge on the above point.[8] My concern is not with the rationality of internalizing a disposition to comply with one's agreements – a decision by necessity made *ex ante* – rather, it is with a gap the conditional cooperation solution fails to address. If I see no benefit to me to enter an agreement (even where I can be assured of mutual cooperation), there is nothing inherent in my new-found disposition to abide by my agreements made that would compel me to enter this agreement as well. Following the discussion in Chapter 5, we would say that while evolution favours those who comply with agreements made, it does not favour making any particular agreement that comes one's way. Otherwise, I would be trapped into entering any offer made to me, including, for example, a request to buy heroin. Thus, my *ex ante* agreement can only be to the rule that I will abide by my agreements made when interacting with like-minded sorts – and not that I will make specific agreements for which I have no *ex ante* way of knowing whether my preferences will track. If I am bound to enter all agreements despite my preferences, my expected utility is negative. The wisdom of such an *ex ante* agreement depends on the total number of distinct interests that are possible (I), divided by the number of distinct interests a given individual has (i) factored by how many players are involved in the potential agreement (n). The odds of a given person having a given interest is the combined probability of $(I/i)^n$. Imagine a world where each individual has merely five distinct interests among a total of only ten possible interests. In this world, imagine two players enter the sphere where an agreement could be made. The odds of mutual benefit, then, is $(5/10)^2 = 0.25$. If the expectation of mutual benefit among fully transparent CCs is only 0.25, it would no longer pay to be a CC: Unconditional Defectors (UDs) would do better. The prospect of mutual advantage being gained between members of a party decrease substantially the more disparate interests there are and the more players are allowed into the game.

There is nothing in the appeal to hypothetically agreeing to a rule to abide by agreements that also impinges on a commitment to make certain agreements. Despite assurances that the other will comply with the agreement so long as we do, it does not follow that in every instance we stand to gain from such compliance. An agreement to play basketball simply does not interest me and I would not agree to it despite assurances from the other players they will not renege on that agreement. In fact, given my preference set, I would do better if they did renege. So it can hardly count that my having internalized a rule to abide by my agreement can be carried over to my also being bound to abide every agreement that is offered to me. Once this is understood, we can interpret requests for aid as offers to enter agreements for which there is no necessary internal motivation to accept – even despite an internalized commitment to be a CC. Nothing in the disposition of being a CC leads me to suppose I have a moral obligation to help those in need. Hence the problem.

8.3. Adjusting the Baseline

At this point, most contractarians argue that it is reasonable for people to agree on some rules in the *ex ante* or hypothetical situation, so long as the hypothetical situation is suitably demarcated. We cannot *ex ante* agree to enter all agreements offered us even with a guarantee of compliance. Still, it may be protested that certain types of agreements can be made *ex ante*, particularly the sort that we would agree to help others out when we can. Given a choice between liberty or rescue, it is not at all obvious that we would prefer liberty. So, although liberty is worth preserving, it is not in any obvious sense a paramount value that we would place above all else. So long as we can reasonably foresee that in an *ex ante* position, reasonable people (intent on finding common moral principles that like-minded individuals could not reasonably reject) would place a value on assistance and frown upon indifference.[9] The fact that we can make such an *ex ante* agreement does not show that it is rational to do so. It might be deemed in your self-interest to make such a hypothetical agreement should it be the case that agreeing to a rule to relieve suffering (when you can) promises rewards to you that outweigh the disadvantages to you. Such a supposition is not obvious, however. The prospect that you will require assistance will surely be outweighed by the prospect of others requiring assistance by you. Even factoring in the benefit to you by your assistance will not obviously outweigh the accumulated costs of helping others. For example, imagine if helping costs you on average two utiles per helping encounter, while being helped benefits you fifty utiles on average per helping encounter. So long as your being required to help outweighs your need to be helped more than 25:1, it would be irrational for you to make an *ex ante* agreement committing yourself to help.

On the other hand, if we already know our lowly position in the world, or if despite being behind a veil of ignorance we at least know the distribution of resources so that we can avail ourselves of the probability of being rich or poor should we choose a non-egalitarian state, we may reasonably presume our requiring help can outweigh the costs of helping. This would not show that it would be reasonable for all to make this decision. We might accommodate this pluralism by committing ourselves to help only those who have likewise committed themselves to help – perhaps easy for virtual players who are transparent, but a murky task for real people. Alternatively, we might alter our conception of a baseline.[10]

It is common to understand the baseline as whatever position one happens to be in prior to the bargaining offer.[11] To reduce an individual to a position lower than his status quo is inadmissible. Nothing is required to elevate an individual from his status quo, however. Mary's being indifferent to Joseph's suffering does not make Joseph's situation worse than he was before Mary's arrival. Joseph is suffering before Mary's arrival, and suffers the same after Mary's departure. Since Mary does not lower Joseph's baseline (defined in terms of the status quo), indifference is permissible. Some may contest that Joseph's suffering is increased, and it is increased precisely because of Mary's indifference. Should Joseph not consent to be made worse off by Mary's indifference, then Mary cannot morally be indifferent according to consent theory. We would have reason to force Mary to not make Joseph worse off, and this is instantiated by forcing Mary to help relieve Joseph's suffering. With only a little reflection we can see the imbecility of such a manoeuvre. That a man deems himself to suffer in the absence of sex is not a sufficient defence for rape. It would be a sufficient defence, note, if we endorse the general rule that refusing a proposal counts as harming the proposer, coupled with an anti-harm rule. As much as we may want a rule telling people to help others when it is feasible, we cannot remove the right to refuse proposals. A plausible suggestion is to alter the definition of a baseline. After all, perpetuating the status quo baseline conception results in wildly unequal distributions. It is precisely because people come to the bargaining table in disproportionate positions that may have propelled contractualists to view the assessment of bargains reached, not from the initial status quo baseline, but rather from some other, fairer, more universal, impartial baseline. We might demand Mary help Joseph only in case Joseph is below his subsistence level. By such reasoning, we harm people if we leave them below their subsistence level. Since a person drowning is below his subsistence level, his request for help cannot be viewed as a mere proposal. Whereas Sally's declining Sam's offer of sex is not a case where Sam is below his subsistence level. By altering the concept of the baseline, we can say without inconsistency that Sally has a right to decline Sam's proposal in a way that Mary has not a right to decline Joseph's proposal. A nice solution, but one open to the Singer and the Loophole problems.

8.4. Singer and Loophole Problems

8.4.1 The Singer Problem

If we ignore national boundaries (as we ought to if we are prescribing a universal moral principle), then the requisite duty to help exceeds our general understanding: billions are presently below their subsistence level right now. Nor can we complain that we are unaware of such suffering given the media's penchant to probe into all corners of the globe and report back to us in truncated, sensationalized productions. As Peter Singer pointed out, the principles we appeal to, to justify our intuitions that we have a duty to help in a particular case (Mary's saving Joseph from drowning, for example), commit us by law of consistency to help out the entire world's population – a task that inevitably requires our selling off much of our possessions and turning it into food, medicine, clothing, and conceivably a military force to rescue people from tyrannical regimes.[12] By the simple recognition of a duty to help a child from drowning in a pond, we are committed to recognize our lives are dreadfully immoral. I shall refer to this as "The Singer Problem."

Singer does not view this as a "problem." He sees in this argument a necessary deduction.[13] I am calling it a "problem" since we now have a choice between accepting the consequent (to drastically alter our way of life) or rejecting the antecedent (to resist any appeal to a moral duty to help relieve suffering), and that both options seem intolerable. If we deny any duty to help relieve suffering, although we escape having to change our comfortable, luxurious, idle, wasteful ways of living, we cannot be morally bound to help the child from drowning. This is the two-pronged fork of the

Singer Problem: it rejects any middle ground solution. We are left with a choice between two intolerable positions: to recognize no duty to help, or to recognize the duty to help exceeds our intuitions.

8.4.2 The Loophole Problem

A further practical problem arises. To endorse a hypothetical contract to relieve suffering, we need to define the duty in such a way that would force us to help in only those situations where we would otherwise be motivated to help prior to our agreement. For on the assurance justification of a duty to help, we are certainly not interested in being bound to do more than we intended. Odysseus agreed to be tied to the mast and have his orders refused; but, again, it was not any order that he wanted refused: only the order to row him ashore to wallow with the Sirens. He would not agree to be whipped while tied and have his orders not to be whipped ignored. Nor would he wish his simple request for water be disavowed.

It is very difficult to imagine how rationality requires our making the *ex ante* agreement to commit ourselves to help without first being able to distinguish between appropriate cases for helping and permissible cases for declining to help. We might (rationally) agree to a duty to help that does not commit us to help when the disutility to the self outweighs the intrinsic benefit accrued to the helper from beneficence itself. Should that distinction be possible, we ought to willingly make such an *ex ante* commitment since it is impossible we will be harmed by it and possible that we will be helped. Unfortunately, such a commitment will be indistinguishable from the absence of such a commitment. The *ex ante* commitment is rendered so loose that it carries no more weight than a subjective post-hoc decision. It effectively admits a loophole for anyone so inclined to take it. Who is most inclined to use the loophole? The answer is the indifferent – precisely the group we intended to target. Admitting loopholes for anyone to take is tantamount to confessing no duty to help exists.

Generally, duties of beneficence are not taken to be "perfect" in the Kantian sense. They are rather "imperfect" duties and require an alternative means of policing.[14] Such talk is problematic, however. To speak of "imperfect" in this sense is misleading. It cannot count as a qualification of a duty. Rather, the qualifier "imperfect" attached to "duty" operates in the same way that "cancelled" affects "performance" in "a cancelled performance." A cancelled performance is not a lesser performance: it is no performance at all.[15] Similarly, an imperfect duty is no duty.

The distinction between duties and ideals will rightly stay with us.[16] Scanlon understands beneficence as requiring a subjective judgement concerning how much help to offer and what counts as an exonerating circumstance.[17] Relying on subjective interpretations will be sufficiently motivating for those already so inclined, but it will fail to motivate precisely those who remain indifferent.[18] No one who is predisposed to indifference will have any motivation to feel that obligation applies to them. To permit loopholes yet scorn all those who avail themselves of such loopholes is evidence of incoherence.[19]

My concern about loopholes is not whether helping can be made law and enforced, but whether or not our condemnation of people availing themselves of loopholes can be justified on amoral grounds. Linking our condemnation to a moral point of view is idle since the moral theorists' task was to explain precisely why the indifferent ought to adopt this moral point of view.

8.5. Natural Affections

The above considerations may have made it clear that beneficence is not something grounded in reason. One might, thereby, view beneficence as a natural starting point. In other words, since we cannot derive duties of beneficence from a bleak Hobbesian state of nature using self-interested reason alone, and since we are creatures prone to benevolence, we must conclude that Hobbes's state of nature is a misguided starting position. Our natural state must include strong passions that, among other things, demand the duty of beneficence.[20]

Since we do agree that helping others is a moral duty of some sort, how ironic that what we agree is moral is not justified by a theory that bases morality on agreement! The complaint that idealized starting points are normatively loaded can now be answered by an appeal to realism. Look, we might say, we are not here to construct a virtual world from bare assumptions for mere interest; rather our goal is to explain the starting point we must have had in order to get here.[21] By adding in the natural or intrinsic utility of helping, the odds may favour a hypothetical agreement to help; an agreement to which CCs are bound to oblige. And this is borne out by the fact that many of us do have a real preference to help which is not derivative on a mere preference to be helped.

Grounding a duty to help on natural affections seems, at first glance, to be superfluous. After all, we all have a natural desire to defecate, yet no duty follows from that. If anything, we have a duty to refrain from defecating (in certain areas). Still, despite our sharing a preference for something, we may require an extra motivation to act on it. It is conceivable that despite our wanting to do x, we know that left to our own devices we probably will not do x. Wanting to exercise is not a sufficient condition for exercising, for example. This is the problem that an *ex ante* agreement attempts to solve.

Notwithstanding our desire to make agreements, we also know that we will be tempted by the lure of profit or sloth to unilaterally renege. Likewise, despite our natural inclinations to help, we may require assurance that our help will (a) not be wasted, and (b) not disadvantage us. Both assurances may be satisfied if we *ex ante* enter a pact with other like-minded individuals similarly thwarted from helping when they desire. If we internally hard-wire ourselves to help, this will certainly increase the odds of successful helping, nor will it disadvantage us, since we are all (or most of us) equally disadvantaged by our generosity.[22]

In the previous chapter, I have already highlighted what the assurance move has achieved: (1) It admits that we have natural sentiments to help. (2) It shows how we may nevertheless require an *ex ante* agreement to commit ourselves to help despite our natural sentiments. (3) The duty comes from being bound to an agreement, and so we have no cause to worry about question-begging or naturalistic fallacies. And (4) the agreement is rationally motivated for all persons who satisfy the two conditions of (a) having natural sympathies to help, and (b) are thwarted from helping voluntarily due to the assurance problem.[23]

Can it work in practice? One difficulty is that it is not known whether we all do have a natural desire to help others. Social-psychological research on bystander intervention indicates that our folk-lore intuitions about this are faulty. For example, studies by Melvin Lerner have shown an interesting effect.[24] Those who tend to believe that the world is a just place to live tend to believe that good things happen to good people and bad things happen to bad people. Such individuals will be less likely to help others in need. After all, if they are in need, they must deserve it. Otherwise, here is a case where a bad thing happens to a good person. This is why Virgil reprimands Dante when Dante's first instinct was to help the unfortunate souls drowning in one of Hell's bolgias. Kicking them back into the mire was the just action, not helping them. This explains the various attribution errors non-helpers tend to employ. Rather than assuming a poor woman collapsed on the sidewalk is a victim of a stroke, it is more convenient to think of her as the victim of her own penchant for drink. Conversely, those less inclined to think the world is naturally just were found to be more inclined to help others. If we can generalize from this, and if we wish to increase helping behaviour in the world, we should become less religious, not more.

Apart from one's belief about the structure of the cosmos, a variety of situational factors can impede helping. For example, the more potential helpers there are lessens the chances of someone's helping.[25] "If no one else is helping, I would appear a fool to help," seems to be the thought process. Alternatively, we may find ourselves preoccupied. Darley and Batson devised a study where they sent seminary students across campus to give a lecture on the

Good Samaritan; their path invariably crossing a man crying out for help. Alas, the seminary students were late for their lectures, and were, after all, preoccupied with fine tuning their inspiring speeches. Few of them stopped.[26]

The social-psychological research noted above does not show that we have no natural sympathies to help, mind you; merely that there are even further obstacles to our acting on these sympathies (if we have them) than merely the assurance problem. Nevertheless, it is within the consent theorist's purview to hypothetically agree on rules that would force us to comply with these rules even when, at the time of being forced to comply, we would rather not. Still, such reasoning does not extend to forcing people to abide by these rules who would not voluntarily agree to them in the *ex ante* position.[27] Refusal is possible so long as there is no natural inclination to help, or that whatever natural inclination exists does not outweigh competing desires. To flatly deny this as a possibility is itself unrealistic. To counter this, some argue that these individuals are missing an intrinsic good.[28] To repeat an argument raised in Chapter 2, even if one really could enjoy being moral for its own sake, this does not exclude the possibility of indoctrination. Saying, "I do good because I am averse to the guilt I feel when I do bad," omits the possibility that the guilt is due to an indoctrination independent of the wrongness of the act. Killing an unsuspecting Ojibway may have seemed a guilt-free act to a cowboy, for example. Masturbation may still seem a guilty pleasure to some. Saying, "I do good for its own intrinsic reward," apart from being psychologically displaced, is also theoretically problematic. If it is the doing good that I enjoy, does it matter whether what I take to be good really is good? And if it makes sense to even ask this question, then we need a criterion for determining whether an act is good independent of one's intrinsic enjoyment of it. Conversely, if the reason one comes to enjoy doing good acts is the goodness of the act, that one enjoys the goodness is irrelevant to our distinguishing whether it is good. In brief, telling Mary she would find utility from helping should she have an intrinsic preference to help is comparable to telling someone who does not like mushrooms that she should like mushrooms on the grounds that she would then be able to derive satisfaction from those mushrooms. However true, such advice is uninteresting. It cannot bind the indifferent. Natural affections alone are insufficient to ground our belief in duties of beneficence.

8.6. The Euthyphro Problem

Although some of us might agree to a helping rule in an *ex ante* position, it is not rationally forced upon us. If we try to accommodate a non-universality, we allow loopholes for the indifferent; precisely our target group. Nothing stops us, however, from asserting that we ought to have extra restrictions on our liberty for independent reasons.[29] Can we impose a duty to enter certain types of agreements independent of any *ex ante* agreement? Of course we can, but the question specific to this chapter is: can a consent theory impose any such further duty? On what grounds can such a pre-agreement duty be founded? For consent theorists, morality is defined by the process of agreement. Here we are demanding that we are morally obliged to make certain types of agreement. In essence, we have stumbled upon the Euthyphro problem. Is it moral because we have agreed to it, or do we agree to it because it is moral?[30] You can see that the latter formulation is death to consent theory. It would show that moral justification comes from somewhere external to the act of agreement, something consent theory denies. And if it is the former, we cannot impose antecedent duties on the sort of agreements that we make. We are forced to choose between tolerating indifference or abandoning consent theory.

If our moral judgements about helping precedes our agreeing about it, then inadequate and convoluted attempts at showing how consent theory might justify the agreement is almost proof enough that our agreement is mandated by our moral presuppositions, and not the reverse. An appeal to an external criterion to force the moral requirement of making an offer to help someone (when they are suffering and the degree of help is feasible) is plainly anathema to consent theory. Proponents of consent theory seem committed to recognize the moral permissibility of being indifferent to the suffering of others. The seeming "immorality" of this conclusion should not escape us, especially in the face of the bloodiest of centuries in the history of the world: the era of indifference.

8.7. The Solution: An Overextended Heuristic

So far the argument in this chapter looks dreary: it is not rational to help others at one's own expense even if one were a CC, although our understanding of morality says we must. But I have taken pains to explain why rationality is not the operative logic concerning morality (Chapter 4), and the above problem has looked on charity from the rational framework. From the rational framework, there is no help. Give it up. But something different emerges from the evolutionary model (Chapter 5). The benefits of altruism are not as straightforward as the development of CC, however. What I shall argue for is that altruism is an overextension of a useful heuristic, and this overextended heuristic has developed into a norm. This may sound unsatisfying, but first let us see how the argument goes. Then I shall try to argue that it is as satisfying as we need.

8.7.1 Moral Norms

In Chapter 5, I argued that success in terms of survival for immoral strategies backfires in terms of heritability. The greater proportion of Defectors in the population means the fitness of Defectors decreases. Strategies that can prosper among their own kind will do better in terms of reproductive advantage than strategies that do not. So long as interactions in nature are more correlated than random, as seems to be the case, evolutionary models of Prisoner's Dilemmas and Ultimatum Games favour moral strategies, not immoral strategies.[31] When speaking of reproductive success, we need not confine ourselves to sexual selection. Cultural beliefs may be inherited without the mediation of gene transference. If an organism has fitness due to a certain behaviour or belief (a phenotype) x, we will say that phenotype x has adaptation. That is, having x is what (or one of the features that) enables that organism to survive. So long as x is also heritable, we can predict the increase of x in that population even when x is not biologically inherited (a genotype). The process of cultural selection is through learning and assimilation. Learning and assimilation are the politically correct terms for the more basic drives of mimicking and habit formation. Mimicking is a built-in strategy that tends to benefit mimickers over non-mimickers.[32] Mimicry comes in three kinds: (1) proximity, (2) prestige, and (3) bandwagon.

Concerning proximity mimicking, we tend to mimic whomever we come across first. This is typically our parents, but it need not be. Mimicking one's parents need not have anything to do with genotypic behaviour, as Konrad Lorenz demonstrated with geese.[33] In terms of moral phenotypes, if our parents engage in certain moral dealings with others, we will more likely engage in similar moral dealings with others. Proximity mimicking will include modelling ourselves after our siblings and, later, our peers. We ape anyone in proximity.

The second kind of mimicry is to mimic the prestigious. This is what Aristotle was recommending, assuming the wise were the prestigious members of the group to which Aristotle identified himself. In prestige imitation, we imitate only the prestigious members of an identified group. These are individuals deemed "successful" by some standard. Coupled with a desire to be successful, we imitate the traits of these figures in the hope that the traits we copy are conducive to the success we want. Sometimes we may model the wrong behaviour, like wearing messy hair because Einstein does when what we want is Einstein's intelligence, or smoking cigarettes because a cool movie star did in a film when what we want is the coolness, not the cancer. An important point to note is that prestige mimicry latches on to anyone we deem prestigious, not necessarily someone who is worthy of mimicry. The prestigious may include the popular kid in school, the Hollywood star, or an Enron CEO, as much as Aristotle's wise man.

The third mimicking ploy is to bandwagon. To bandwagon is to mimic what the majority does. Neighbours share gardening habits. Disrepair breeds disrepair. Topiary breeds topiary. Individualists scoff. What, no originality, no autonomy? But according to moral conventionalism, it is precisely this tendency to conform in humans which generates moral conventions. Advertisers, political propagandists, and social psychologists are well aware of how much bandwagoning moves us – even those of us who profess otherwise. Studies by Solomon Asch offer ample illustration. In one case, participants are asked to identify in public which of three lines are longer, when it is clearly the case that line *A* is longer than lines *B* or *C*. Few in such cases have the courage to pick *A* when all the other "participants" – confederates of Asch – have already publicly picked *C*.[34] Other studies reveal that if you get into an elevator where everyone is facing the back wall, you too will likely face the back wall. We are conformers, whether we like it or not.

The three different strategies of mimicking behaviour often conflict. Should one mimic one's parents when they are not prestigious? Should one mimic the prestigious when they are not popular? Should one mimic the popular when one's parents disapprove? For Aristotle, mimicking the wise was obviously superior to, and in conflict with, bandwagoning and proximity mimicry. A developmental model lends support to Aristotle's contention, since prestige mimicry requires higher developmental sophistication than other forms of mimicry. Mimicking one's parents requires little discrimination; mimicking the popular requires a little bit more; and mimicking the prestigious requires the most amount of discrimination. Assuming whatever is developmentally later is superior is not always a correct assumption, however. The mature sea squirt, for example, is developmentally inferior to the immature sea squirt,[35] and much poetry laments the change from childhood to staid, responsible, maturity.

Picking prestige mimicry as the superior form of mimicry, however, makes the mistake of assuming a competition among the mimicking strategies. That is the wrong way of looking at the mimicking process. The three forms are interrelated. Proximity determines which group one will identify with. Bandwagoning determines who counts as successful within that group. Mimicking the prestigious is only the last phase of the mimicking strategy. The prestigious are identified with success.[36] Individuals are driven to succeed, and so a good strategy is to mimic those who are successful. Since notions of success vary from culture to culture, individuals deemed prestigious will vary from one group to another. A prestigious hoodlum may not influence someone who does not feel any belonging to the group of hoodlums. A prestigious businessman may not interest a wannabe philosopher. Whom we deem prestigious, therefore, is already dependent on the group with which we identify. The notion of mimicking the prestigious does not presuppose any objective criterion for who counts as prestigious. Aristotle would not have been happy about this. He was aware of the relativism in identifying the wise. Rather than suggesting that the wise is a relativistic notion, he thought merely that most people just got it wrong. He explained this in the following terms: those on the excess side of the mean will deem the wise man as someone suffering from defect, whereas those on the defect side of the mean will deem the wise man suffers from excess.[37] The result: only the wise can recognize who the wise are: a nice non-falsifiable comeback to anyone who questions the actions of the wise. Boyd and Richerson's model, on the other hand, takes the mystery out of the process of identifying the wise. The wise are a subset of the prestigious members of a group, and prestige is defined by success, and success is defined in terms of convention.[38]

Cultural differences do not affect the conventionalist picture. The best strategy is to mimic the successful. Merely, the concept of success will vary from culture to culture and group to group. Who counts as "wise" is a conventionalist notion, just as which virtues the wise prioritize will vary among cultures.[39]

Doing as the wise do will not help as *moral* advice unless the wise also are moral. Since the definition of the wise varies from culture to culture, the chances of their *all* being moral seems slight. At any rate, it is unlikely that they are deemed successful *because* they are moral. Some may be deemed successful precisely for their *immorality*. On the moral conventionalist model, morality has to do with our relations with others.[40] One's dealing with others successfully has a lot to do with one's reputation concerning how one has dealt with others in the past. If we understand fair dealing with others as moral dealing, and reputation has to do with fair dealing, then the greater need for preserving one's reputation will be positively correlated with moral behaviour. In Gauthier's defence of translucency – the ability to accurately detect others' dispositions a significant amount of time – he relied on precisely this rationale for making oneself transparent.[41] On the evolutionary account, moral action is a payment for preserving your reputation. If this were the impetus of moral behaviour, those more in the limelight would have a greater need for being recognized as moral players. Since those identified as the prestigious are certainly more in the limelight, the prestigious will have more need for moral action than the plebeian. If the plebeian knew the mechanics of morality, they would be less moral, but fortunately, evolutionary forces mandate they follow the heuristic of mimicking the most successful. Since the most successful tend to be moral (due to the limelight and the need to preserve their reputation), the plebeians tend to copy that moral behaviour as well. This may be a poor move rationally speaking, but it does have evolutionary advantage. By this analysis, agents are moral for the wrong reason, but since moral dealing has evolutionary fit, the folly will spread.

Not all beneficial social behaviours that evolve, evolve for those beneficial reasons. On the wrong belief that malaria was caused by bad air (*mal aria*), a preventive cure was to shut the windows. This had the fortunate side effect of keeping the infected mosquitoes out, and so the successful convention of shutting windows passed through the generations.[42] Mimicking the wise may likewise have the unintended happy consequence of driving moral norms to fixation. Saying this does not mean that moral norms are false. What is false in the malaria case is that malaria is spread by bad air. What is false in the moral case is that we have a categorical duty to help those in need. What is true in the malaria case is that shutting the window is a good strategy to prevent malaria. What is true in the moral case is that behaving in ways that signal to others you are reliable cooperator has evolutionary fitness, and that helping others when they are in need is a good way of signalling to others that you are a reliable cooperator. None of this need work on the level of rational motivation.

Summary. Imitation and social punishment are cultural ways of stabilizing norms. But why these norms? Any norms can be stabilized if imitated while deviants are punished. But imitation need not be random. If imitation is predicated on success, then successful strategies will tend to rise to the norm. Research by Boyd and Richerson, for example, make exactly that case.

8.7.2 Altruism

When I speak of altruism in what follows, I mean the moral tendency found in humans to help even unrelated fellow humans. I do not wish to define it so narrowly that it necessarily excludes the prospects for mutual advantage. Unlike market transactions, however, which also have the effect of helping others, the intention in altruistic acts is not primarily self-interested. So far, we have grounds for thinking that social norms tend to be profitable for the individuals in that group on the basis that norms tend to coagulate around success, and success is predicated on empirical features of the environment. By itself, this does not tell us that altruism is one of those norms. The mere existence of altruism and the social practice of condemning egoism at a time when altruism would be expected, coupled with the tendency that norms coagulate around success, gives us prima facie reason to suspect that altruism has fit, despite its individual costs.

So far the argument is only speculative. From the imitation argument in which social norms are developed through imitation of successful phenotypes, and from the fact that norms governing altruism exist, we infer that altruism likely has evolutionary benefits, or is associated with a successful strategy. But not everything that was fit necessarily remains fit, or remains fit for the same reason as its historical fitness. The success of altruism, at any rate, is not demonstrated by the games examined in Chapter 5. Those games demonstrate the success of conditional agents. Altruism is an unconditional strategy. So the mere inference that altruism must have evolutionary fit given that it is supported by a current norm is hardly convincing. But we can argue that altruism is associated with evolutionary benefits. That is, the norm of altruism piggy-backs on a successful strategy. In this sense, we can explain the prevailing norm of altruism as an overextended heuristic.

If altruism is to be culturally enforced, we cannot ignore the loophole problem (see 8.5). What this tells us is that the environmental cues for when altruism is to be expected are fuzzy.[43] Demanding fine discriminating skills in knowing when altruism is expected and when not is inconsistent with the imitation model. Imitation is a coarse-grained operative. When imitating the successful, we are not quite clear on what features we should imitate. It is unlikely that we will stumble upon exactly the right features at exactly the right degree. Therefore, we can expect to err. The question is, do we err by not copying enough, or by copying too much? Boyd and Richerson pitted insufficient imitation strategies against overextended imitation strategies and found that the redundant imitators outdid the insufficient imitators in terms of success.[44] In fact, there will be a tendency toward imitation in greater degrees. Individuals attempt to discover the best behaviour in varying environments. They may learn on their own, or they may imitate those of previous generations. Given the vagaries of such experimentation, varying conditions and third variables, strategies that are effective in the short term are not necessarily strategies that succeed in the long haul. A risk of self-experimentation is latching onto behaviours that have no lasting effect. Imitating successful members of previous generations, therefore, has the benefit of avoiding costly errors. The risk of imitation is greater when environments have shifted. Of course, there is a wide range between full imitation and zero imitation. So long as environments change, a need for self-experimentation will always be useful. Cases where self-experimenters succeed, however, become a case for imitation of subsequent generations. Therefore, for each successive generation, a stronger case can be made for the imitation strategy. Assume the average strategy is to imitate $x\%$

of the time. Invaders can vary in terms of their imitation strategies, some greater than average, some less. Assuming relatively stable environments, the greater imitators will achieve a higher payoff than both the average and the lesser imitators. As Boyd and Richerson observe, "when it is difficult for an individual to determine the best behavior and when environments change infrequently, more than 90 % of a population at equilibrium simply copies the behaviour of others."[45]

Assuming heuristics are not perfectly sensitive to all environmental cues, we can expect some of the adaptive heuristics may be overextended in certain cases. As long as such errors have no maladaptive effects, we can predict the norm to include the overextension over future generations. This would explain why norms of altruism become entrenched so long as the prestige members are altruistic in the first place, or behave in ways that may appear altruistic to potential imitators. But why would they be altruistic? Why would successful progenitors have altruistic tendencies? So far, all the argument has suggested is that if altruism is a successful strategy, the drive toward mimicry and conformity would preserve altruism as a norm. But why would the norm of altruism have evolved in the first place?

An added feature of the overextended heuristic account may be offered. *A*'s failing to be altruistic may signal to a conditional cooperator *B* that *A* may not be a CC, after all. This may be false, since, as we have described above, a CC need not commit to every proposal that comes his way. Earlier (4.3), we discussed the prospect of detection errors. At such time, we limited the discussion to only those mistakes where a CC incorrectly takes a UD to be a CC (or UC). But mistakes can work the opposite way. A CC may mistake another CC for a UD. CC's conditional response means he must be sensitive to some cues, however broad-brushed they may be. The failure of a CC being altruistic – even in cases where altruism is only fuzzily suggested – can be enough of a signal to potential cooperators to read the non-altruist as a non-CC. If this norm were itself fostered through imitation and social censure, then any CC would be motivated to be altruistic within the realm of the fuzzy norm. That is, CCs will be moved to be charitable for the purposes of maintaining one's reputation for being a CC. Such theorizing may also support why society currently treats CCs as being preferable to RCs, since CCs will, while RCs will not, cooperate with UC – an act of charity, after all.

This line of argument carries more weight for those whose reputations are more prominent. Prestigious members will be more motivated than less prestigious members to guard their cooperative reputation, if merely to thwart misconstrual, and thus defection, by others. To preserve their reputation as a CC, therefore, they may behave in ways technically unnecessary for conditional cooperation. The fear of a marred reputation moves the prestigious toward overextending cooperative behaviours into fuzzy cases of altruism. Meanwhile, the broad-stroked mechanics of mimicry move the plebeians to imitate the overextended altruism of the prestigious, which in turn coagulates into a norm.

This account will not vindicate Singer, by any means. Some may complain mimicry and social norms are entirely too conservative a model. My point here is far less lofty. I offer an account of our moral attitudes toward altruism from the vantage of an evolution-backed consent theory. Notice, too, this is not Joyce's fictive stance.[46] At best, it is a partial fictive stance. There is nothing fictitious about the benefits of the norm, merely about the moral justification typically offered in support of the norm.

Notes

[1] For example, David O. Brink, "Self-Love and Altruism," in Ellen Frankel Paul, Fred Miller, Jr., and Jeffrey Paul (eds.) *Self-Interest* (Cambridge: Cambridge University Press, 1997), 123. See also, his "Rational Egoism, Self, and Others," in O. Flanagan and A. Rorty (eds.) *Identity, Character, and Morality* (Cambridge, Mass.: MIT Press, 1990). See also Jody S. Kraus, *The Limits of Hobbesian Contractarianism* (Cambridge: Cambridge University Press, 1993), 38–9, 45.

[2] John Rawls, *A Theory of Justice* (Cambridge, Mass.: Harvard University Press, 1971). Kavka uses a thinner veil of ignorance in his more Hobbesian contractarian formulation. Gregory Kavka, *Hobbesian Moral and Political Theory* (Princeton: Princeton University Press, 1986), 194.

[3] Thomas Scanlon, *What We Owe to Each Other* (Cambridge, Mass.: Harvard University Press, 1998), 192, 224. In *A Theory of Justice*, Rawls restricts his agents to those situated behind a veil of ignorance and argues that they would select a maximin rule (Rawls, 152–7). The maximin rule is better fitted to a zero-sum game, however. In *Political Liberalism*, Rawls adjusts his concept more along the lines of Scanlon: the "reasonable" are those who seek "fair terms of cooperation." John Rawls, *Political Liberalism* (New York: Columbia University Press, 1993), 49. Similarly, Kavka argues that idealized agents in a hypothetical situation (ignorant of their social positions) would seek to make agreements concerning social goods in ways no other idealized agent could rationally reject (Kavka, *Hobbesian Moral and Political Theory*, 401). Kavka argues that since ideally rational agents would make such an agreement, it is rational for real agents to do so as well (Kavka, *Hobbesian Moral and Political Theory*, 399). So long as rational acts are dependent on context, situation, and information, however, this is far from obvious. Another defense of a restricted understanding of a rational agreement is given by Hare. A rational act is one that satisfies the preferences an agent would have should she be purged of logical error and is sufficiently informed of relevant facts. H.M. Hare, *Moral Thinking* (Oxford: Oxford University Press, 1981), 105–6.

[4] Consider, for example, the Dating Game (or the Battle of the Sexes). See Duncan Luce and Howard Raiffa, *Games and Decisions* (New York: Dover Publications, 1985), 90–4.

[5] David Gauthier, *Morals By Agreement* (Oxford: Clarendon Press, 1986), 167–70.

[6] We may try to conceive Gauthier's Lockean Proviso as an institutionalized response to distress. The Lockean Proviso rules out the taking advantage of another (Gauthier, *Morals By Agreement*, 205–6), but mere indifference cannot count as taking advantage of someone.

[7] We may also believe that certain types of agreements ought to be forbidden, and thus even proposing them ought to be squelched. See Malcolm Murray, "Lewd Propositions and Contractarianism," in Kenneth Cust (ed.) *20th Century Values* (Value-Net, 2002) at http://value-net.org/Publications/20thCentury_Values/20thcentury_values.html.

[8] See Gauthier, *Morals By Agreement*, 9.
[9] Apart from Rawls, Scanlon, and Kavka already discussed, see Gerald Dworkin, *The Theory and Practice of Autonomy* (Cambridge: Cambridge University Press, 1988), 104–8; Ronald Dworkin, *A Matter of Principle* (London: Harvard University Press, 1985), 228–32; Will Kymlicka, *Liberalism, Community, and Culture* (Oxford: Clarendon Press, 1989), 16–9; Margaret Moore, *Foundations of Liberalism* (Oxford: Clarendon Press, 1993), 188–98, and Joseph Raz, *The Morality of Freedom* (Oxford: Oxford University Press, 1988), 407–29.
[10] Rawls's use of the veil of ignorance is to do just this (Rawls, *Justice*, 136).
[11] For example, see Robert Nozick, *Anarchy State and Utopia* (New York: Basic Books, 1971), 84. See also Gauthier, *Morals By Agreement*, 204.
[12] Peter Singer, "Famine, Affluence, and Morality," *Philosophy and Public Affairs* 1 (1972): 229–43.
[13] Singer assumes two principles: 1. Suffering from lack of food, shelter, and medical care is bad; and 2. If it is in our power to prevent something bad (or very bad) from happening, without thereby sacrificing anything of comparable moral importance, we ought, morally, to do it. Singer claims that if his two principles were acted upon, "our lives, our society, and our world would be fundamentally changed" (Singer, 230).
[14] Immanuel Kant, *Grounding for the Metaphysics of Morals* (James Ellington (trans.), Indianapolis: Hackett Publishing, 1981), 30, nt. 12 [Ak. 421], and 32 [Ak. 424]. Mill, too, spoke of obligations we have that, although not punishable by law for omission, are open to social censure: "punishable by opinion, though not by law." J.S. Mill, *On Liberty* (Hackett Publishing, [1859] 1978), ch. IV, 73.
[15] Nebojsa Kujundzic, "On the Logic of Adjectives," *Dialogue: Canadian Philosophical Review* 40/Fall (2001): 803–9.
[16] For a clear discussion on the distinction, see Bernard Gert, *Morality: Its Nature and Justification* (Oxford: Oxford University Press, 1998), 247.
[17] Scanlon, *What We Owe to Each Other*, 225.
[18] On Scanlon's behalf, the indifferent cannot reasonably reject a proposal of an imperfect duty, since, after all, an imperfect duty still permits indifference if indifference is subjectively determined to be appropriate. Thus, for those intent on discovering universalizable principles, an appeal to imperfect duties should work. My point here, however, is that such "success" is hollow.
[19] For Kant, we are not culpable for failing to fulfil our imperfect duties, although it does reveal "a deficiency in moral worth," (*The Metaphysics of Morals* (Mary Gregor (trans.), Cambridge: Cambridge University Press, 1998), 153 [Ak. 390–1]). See Thomas Hill, Jr. for an extended discussion. T. Hill, Jr., *Dignity and Practical Reason in Kant's Moral Theory* (Ithaca, N.Y.: Cornell University Press, 1992), ch. 8.
[20] A biological explanation may also be given. See, for example, Brian Skyrms, *Evolution of the Social Contract* (Cambridge: Cambridge University Press, 1996), 45–62. Gauthier speaks of the necessity to internalize our desire to be a CC to ensure personal compliance even in the single-shot PD. Finding intrinsic worth in cooperation does not simply equate to agreeing to a general rule of beneficence, however.
[21] For a nice account of the difference between two traditions, see Skyrms, *Evolution of the Social Contract*, ix.
[22] We are not disadvantaged relative to others. Helping will typically disadvantage us in non-relativistic terms.
[23] The helping dilemma is then portrayed as a coordination problem, not a prisoner's dilemma.
[24] Melvin Lerner and D.J. Miller, "Just World Research and the Attribution Process: Looking Back and Ahead," *Psychological Bulletin* 85/5 (1978), 1030–51.
[25] See for example, J.M. Darley and B. Latané, "Bystander Intervention in Emergencies: Diffusion of Responsibility," *Journal of Personality and Social Psychology* 8 (1968): 377–83. B. Latané and J.M. Darley, *The Unresponsive Bystander: Why Does He Not Help?* (Englewood Cliffs: Prentice Hall, 1970).
[26] J.M. Darley and C.D. Batson, "From Jerusalem to Jericho: A Study of Situational and Dispositional Variables in Helping Behavior," *Journal of Personality and Social Psychology* 27 (1973): 100–8.

[27] For example, see Jan Narveson, *The Libertarian Idea* (Philadelphia: Temple University Press, 1988).
[28] Aside from Aristotelean doctrines (for example, Susan Wolf, "Happiness and Meaning: Two Aspects of the Good Life," in Ellen Frankel Paul, Fred Miller, Jr., and Jeffrey Paul (eds.) *Self-Interest* (Cambridge: Cambridge University Press, 1997), 207–5, and David Brink, "Self-Love and Altruism," 122–57), see Gregory Kavka, "A Reconciliation Project," in D. Copp and D. Zimmerman (eds.) *Morality, Reason, and Truth* (Lanham, Md.: Rowman and Allenheld, 1984), 297–319: "[T]here are special significant pleasures or satisfactions that accompany regular moral action and the practice of a moral way of life that are not available to (unreformed) immoralists and others of their ilk."
[29] In *A Theory of Justice*, Rawls proposed a two-tiered policy: (1) Everyone has a moral duty not to do anything to others without their consent, and (2) Everyone has a further moral duty to help relieve suffering when possible (Rawls, *Justice*, 60). To avoid the Singer Problem (where we are forced to help others by reducing ourselves to a point merely a fraction above our subsistence level), we can assert, like Rawls, that where conflicts arise between (1) and (2), (1) always trumps (2) (Rawls, *Justice*, 250). Unfortunately, since society's forcing me to do *x* is *prima facie* a violation of my consent, it appears (2) inevitably conflicts with (1), and hence the priority rule provides a loophole against our (putative) moral obligation to help. See for example Jan Narveson, "A Puzzle About Economic Justice in Rawls' Theory" (*Social Theory and Practice* 4/1 (1976): 1–28, and H.L.A Hart, "Rawls on Liberty and Its Priority" *University of Chicago Law Review* 40 (1973): 535–55. In *Political Liberalism*, Rawls fixes this by weakening the first principle: "Each person has an equal claim to a fully *adequate* scheme of equal basic rights and liberties, which scheme is compatible with the same scheme for all; and in this scheme the equal political liberties, and only those political liberties, are to be guaranteed their *fair* value" (Rawls, *Political Liberalism*, 5, my emphasis).
[30] Judith Thompson also raises this conundrum. "[I]t is the patent wrongfulness of the conduct that explains why there would be general agreement to disallow it." *The Realm of Rights* (Cambridge, Mass.: Harvard University Press, 1990), 30, nt.19.
[31] Skyrms, *Evolution of the Social Contract*, 63–79.
[32] Luca Luigi Cavalli-Sforza and Marcus Feldman, *Cultural Transmission and Evolution* (Princeton: Princeton University Press, 1981). Dan Sperba, *Explaining Culture: A Naturalistic Approach* (Oxford: Oxford University Press, 1996).
[33] Konrad Lorenz, *Studies in Animal and Human Behavior*, Robert Martin (trans.) (Cambridge: Cambridge University Press, 1970). This process is referred to as imprinting.
[34] Solomon Asch, "Studies of Independence and Conformity: A Minority of One Against a Unanimous Majority," *Psychological Monographs*, 70/9 (1956): Whole No. 416, Solomon Asch, "Effects of Group Pressure upon the Modification and Distortion of Judgement," in M.H. Guetzkow (ed.) *Groups, Leadership, and Men* (Pittsburgh: Carnegie, 1951), 117–90.
[35] Owen Flanagan, *Varieties of Moral Personality: Ethics and Psychological Realism* (Cambridge, Mass.: Harvard University Press, 1991), 192–3.
[36] Robert Boyd and Peter Richerson, *Culture and the Evolutionary Process* (Chicago: University of Chicago Press, 1985).
[37] "Hence also the people at the extremes push the intermediate man each over to the other, and the brave man is called rash by the coward, cowardly by the rash man, and correspondingly in the other cases." Aristotle, *Nicomachean Ethics*, bk. II. ch. 8 [1108^{b}] 962.
[38] This view is reminiscent of Richard Rorty, *Philosophy and the Mirror of Nature* (Princeton: Princeton University Press, 1979), but cultural evolutionists need not be anti-realists in the Rorty sense. Rorty begins from Hume's impossibility to know reality outside our skins and concludes that therefore reality talk is senseless. But the conclusion does not follow from the premises. For more on this, see Hilary Putnam, *The Collapse of the Fact/Value Dichotomy and other Essays* (Cambridge, Mass.: Harvard University Press, 1992), 100.
[39] Alasdair MacIntyre, *After Virtue* (Notre Dame: University of Notre Dame Press, 1984), 181–5. Admittedly, MacIntyre sides with Aristotle's thesis.
[40] The concept of well-being and morality are not equivalent, obviously, even though one may argue that well-being also has something to do with relations with others.
[41] Gauthier, *Morals By Agreement*, 174–7.

[42] Joe Heinrich, Wulf Albers, Robert Boyd, Gerd Gigerenzer, Kevin McCabe, Axel Ockenfels, and Peyton Young, "What is the Role of Culture in Bounded Rationality?" in Gerd Gigerenzer and R. Selten (eds.) *Bounded Rationality: The Adaptive Toolbox* (Cambridge, Mass.: MIT Press, 2002), 355.
[43] For a nice account of the merits of fuzziness in moral thinking, see Richmond Campbell and Jennifer Woodrow, "Why Moore's Open Question is Open: The Evolution of Moral Supervenience," *The Journal of Value Inquiry* 37 (2003), particularly the discussion on 356–7.
[44] Robert Boyd and Peter J. Richerson, "Norms and Bounded Rationality," in Gerd Gigerenzer and Reinhard Selten (eds.) *Bounded Rationality: The Adaptive Toolbox* (Cambridge, Mass.: MIT Press, 2002), 281–96.
[45] Boyd and Richerson, *Bounded Rationality*, 286.
[46] Richard Joyce, *The Myth of Morality* (Cambridge: Cambridge University Press, 2001).

BIBLIOGRAPHY

Alexander, Jason. "Group Dynamics in the State of Nature," *Erkenntnis* 55 (2001): 169–82.

Aristotle. "Metaphysics," in Richard McKeon (ed.), *The Basic Works of Aristotle* (New York: Random House, 1941).

Aristotle. "Nicomachean Ethics," in Richard McKeon (ed.), *The Basic Works of Aristotle* (New York: Random House, 1941).

Asch, Solomon. "Effects of Group Pressure upon the Modification and Distortion of Judgement," in M.H. Guetzkow (ed.), *Groups, Leadership, and Men* (Pittsburgh: Carnegie, 1951), 117–90.

Asch, Solomon. "Studies of Independence and Conformity: A Minority of One Against a Unanimous Majority," *Psychological Monographs* 70/9 (1956): Whole No. 416.

Aumann, Robert. "Subjectivity and Correlation in Randomized Strategies," *Journal of Mathematical Economics* 1 (1974): 67–96.

Aumann, Robert. "Correlated Equilibrium as an Expression of Bayesian Rationality," *Econometrica* 55 (1981): 1–18.

Axelrod, Robert. *The Evolution of Cooperation* (New York: Basic Books, 1984).

Ayer, A.J. *Language, Truth, and Logic* (New York: Dover Publications, 1952).

Baier, Kurt. "Rationality, Value, and Preference," *Social Philosophy and Policy* 5/2 (1988): 17–45.

Baier, Kurt. "Moral Reasons and Reasons to be Moral," in A.I. Goldman and J. Kim (eds.), *Values and Morals* (Dordrecht: D. Reidel, 1978).

Becker, Lawrence. "The Finality of Moral Judgments: A Reply to Mrs. Foot," *The Philosophical Review* 82/3 (1973): 364–70.

Bentham, Jeremy. in J.H. Burns and H.L.A. Hart (eds.), *An Introduction to the Principles of Morals and Legislation* (New York: Oxford University Press, 1970).

Binmore, Ken. *Game Theory and the Social Contract: Vol. 1: Playing Fair* (Cambridge, Mass.: MIT Press, 1998).

Blackburn, Simon. *Ruling Passions* (Oxford: Oxford University Press, 2000).

Blackburn, Simon. "Moral Realism," in J. Casey (ed.), *Morality and Moral Reasoning* (London: Methuen, 1971).

Bloomfield, Paul. *Moral Reality* (Oxford: Oxford University Press, 2001).

Bouwsma, O.K. "Descartes's Evil Genius," *The Philosophical Review* 65 (1949): 141–51.

Boyd, Robert and Peter Richerson. "Norms of Bounded Rationality," in Gerd Gigerenzer and Reinhard Selten (eds.), *Bounded Rationality: The Adaptive Toolbox* (Cambridge, Mass.: MIT Press, 2002).

Boyd, Robert and Peter Richerson. *Culture and the Evolutionary Process* (Chicago: University of Chicago Press, 1985).

Brandt, Richard. *Facts, Values, and Morality* (Cambridge: Cambridge University Press, 1996).

Brandt, Richard. *A Theory of the Good and the Right* (Amherst, N.Y.: Prometheus Books, 1998).

Brink, David. *Moral Realism and the Foundations of Ethics* (Cambridge: Cambridge University Press, 1989).

Brink, David. "Rational Egoism, Self, and Others," in O. Flanagan and A. Rorty (eds.), *Identity, Character, and Morality* (Cambridge, Mass.: MIT Press, 1990).

Brink, David. "Self-Love and Altruism," in E. Frankel Paul, F. Miller, Jr., and J. Paul (eds.), *Self-Interest* (Cambridge: Cambridge University Press, 1997), 122–57.

Butler, Joseph. "Upon Human Nature," from *Fifteen Sermons Preached at the Rolls Chapel* (London, 1726). Reprinted in Bernard Baumrin (ed.), *Hobbes's Leviathan: Interpretation and Criticism* (Belmont, Calif.: Wadsworth Publishing Co., 1969).

Campbell, Richmond. "Background for the Uninitiated," in R. Campbell and L. Sowden (eds.), *Paradoxes of Rationality and Co-operation* (Vancouver: University of British Columbia Press, 1985).

Campbell, Richmond and Jennifer Woodrow. "Why Moore's Open Question is Open: The Evolution of Moral Supervenience," *The Journal of Value Inquiry* 37 (2003): 353–72.

Casebeer, William. *Natural Ethical Facts: Evolution, Connectionism, and Moral Cognition* (Cambridge, Mass.: MIT Press, 2005).

Cavalli-Sforza, Luca Luigi, and Marcus Feldman. *Cultural Transmission and Evolution* (Princeton: Princeton University Press, 1981).

Churchland, Paul. "Toward a Cognitive Neurobiology of the Moral Virtues," *Topoi* 17 (1998): 83–96.

Copp, David. *Morality, Normativity, and Society* (Oxford: Oxford University Press, 1995).

Copp, David. "Moral Knowledge in a Society Centered Moral Theory," in Walter Sinnott-Armstrong and Mark Timmons (eds.), *Moral Knowledge? New Essays in Moral Epistemology* (Oxford: Oxford University Press, 1996), 243–66.

Crisp, Roger and Brad Hooker (eds.). *Well-Being and Morality: Essays in Honour of James Griffin* (Oxford: Oxford University Press, 2000).

Danielson, Peter. *Artificial Morality: Virtuous Robots for Virtual Games* (London: Routledge, 1992).

Danielson, Peter. "Evolutionary Models of Co-operative Mechanisms: Artificial Morality and Genetic Programming," in Peter Danielson (ed.), *Modeling Rationality, Morality, and Evolution* (Oxford: Oxford University Press, 1989), 423–41.

Danielson, Peter (ed.). *Modeling Rationality, Morality, and Evolution* (Oxford: Oxford University Press, 1998).

Darley, J.M. and B. Latané. "Bystander Intervention in Emergencies: Diffusion of Responsibility," *Journal of Personality and Social Psychology* 8 (1968): 377–83.

Darley, J.M. and C.D. Batson. "From Jerusalem to Jericho: A Study of Situational and Dispositional Variables in Helping Behavior," *Journal of Personality and Social Psychology* 27 (1973): 100–8.

Darwall, Stephen. "Self-Interest and Self-Concern," in E. Frankel Paul, F. Miller, Jr., and J. Paul (eds.), *Self-Interest* (Cambridge: Cambridge University Press, 1997), 158–78.

Dawes, R.M. *Rational Choice in an Uncertain World* (Fort Worth: Harcourt Brace College Publishers, 1988).

Dawkins, Richard. *The Selfish Gene* (Oxford: Oxford University Press, 1976).

Dawkins, Richard. *The Blind Watchmaker* (Hammondsworth: Penguin, 1991).

de Jasay, Anthony. *Social Contract, Free Ride: A Study of the Public Goods Problem* (Oxford: Oxford University Press, 1990).

Dimock, Susan. "Defending Non-Tuism," *Canadian Journal of Philosophy* 29/2 (1999): 251–74.

Dumouchel, Paul. "Rational Deception," in Caroline Gerschlager (ed.), *Deception in Markets* (New York: Palgrave Macmillan, 2005), ch. 2.

Dworkin, Gerald. *The Theory and Practice of Autonomy* (Cambridge: Cambridge University Press, 1988).

Dworkin, Ronald. "The Original Position," in N. Daniels (ed.), *Reading Rawls* (New York: Basic Books, 1976).

Dworkin, Ronald. *A Matter of Principle* (Cambridge, Mass.: Harvard University Press, 1985).

Epictetus. "Encheiridion," in W.A. Oldfather (trans.), *Epictetus: Vol. II* (Cambridge, Mass.: Loeb Classical Library, Harvard University Press, 1996).

Feinberg, Joel. *Freedom and Fulfillment: Philosophical Essays* (Princeton: Princeton University Press, 1992).

Feinberg, Joel. "Wrongful Life and the Counterfactual Element in Harming," in Joel Feinberg, *Freedom and Fulfillment: Philosophical Essays* (Princeton: Princeton University Press, 1992), 3–36.

Fishkin, James. "Bargaining, Justice, and Justification: Towards Reconstruction," *Social Philosophy and Policy* 5/2 (1988): 46–64.

Flanagan, Owen and A. Rorty (eds.). *Identity, Character, and Morality* (Cambridge, Mass.: MIT Press, 1990).

Flanagan, Owen. *The Varieties of Moral Personality: Ethics and Psychological Realism* (Cambridge, Mass.: Harvard University Press, 1991).

Foot, Philippa. "Morality as a System of Hypothetical Imperatives," *Philosophical Review* 71 (1972): 305–16.

Foot, Philippa. *Natural Goodness* (Oxford: Oxford University Press, 2001).

Foot, Philippa. *Virtues and Vices* (Oxford: Oxford University Press, 2002).

Frankfurt, Harry. "Freedom of the Will and the Concept of a Person," in H. Frankfurt, *The Importance of What we Care About* (Cambridge: Cambridge University Press, 1998), 11–25. Originally published in *The Journal of Philosophy* 68/1 (1971).

Frege, Gottlob. *The Foundations of Arithmetic*, J.L. Austin (trans.) (Evanston, IL: Northwestern University Press, [1884] 1953).

Gass, William. "The Case of the Obliging Stranger," *The Philosophical Review* 66 (1957): 193–204.

Gauthier, David. *Morals by Agreement* (Oxford: Oxford University Press, 1986).

Gauthier, David. "Moral Artifice," *Canadian Journal of Philosophy* 18 (1988): 385–418.

Gauthier, David. "Morality, Rational Choice, Semantic Representation: A Reply to My Critics," *Social Philosophy and Policy* 5/2 (1988): 173–221.

Gauthier, David. *Moral Dealing: Contract, Ethics, and Reason* (Ithaca and London: Cornell University Press, 1990).

Gauthier, David. "Rational Constraint: Some Last Words," in Peter Vallentyne (ed.), *Contractarianism and Rational Choice* (Cambridge: Cambridge University Press, 1991), 323–30.

Gauthier, David. "Why Contractarianism?" in Peter Vallentyne (ed.), *Contractarianism and Rational Choice* (Cambridge: Cambridge University Press, 1991), 15–30.

Gauthier, David. "Assure and Threaten," *Ethics* 104 (1994): 690–721.

Geach, P.T. "Good and Evil," *Analysis* 17 (1956): 33–42.

Gert, Bernard. *Morality: Its Nature and Justification* (Oxford: Oxford University Press, 1998).

Gewirth, Alan. "Can Any Final Ends Be Rational?" *Ethics* 102 (1991): 66–95.

Gibbard, Allan. *Wise Choices, Apt Feelings: A Theory of Normative Judgment* (Cambridge, Mass.: Harvard University Press, 1990).

Gigerenzer, Gerd and Reinhard Selten. "Rethinking Rationality," in Gerd Gigerenzer and R. Selten (eds.), *Bounded Rationality: The Adaptive Toolbox* (Cambridge, Mass.: MIT Press, 2002), 1–11.

Gilbert, Alan. *Democratic Individuality* (Cambridge: Cambridge University Press, 1990).

Glasser, William. *Reality Therapy: A New Approach to Psychiatry* (New York: Harper and Row, 1965).

Goldman, A.I. and J. Kim (eds.). *Values and Morals* (Dordrecht: D. Reidel, 1978).

Goldstick, Daniel. "Objective Interests," in H. Lyman, John Legters, P. Burke, and Arthur DiQuattro (eds.), *Critical Perspectives on Democracy* (Lanham, Md.: Roman & Littlefield, 1994), 147–64.

Goodin, R. and P. Pettit (eds.). *A Companion to Contemporary Political Philosophy* (Oxford: Blackwell Publishers, 1993).

Gould, Stephen Jay and Richard Lewontin. "The Spandrels of San Marco and the Panglossian Paradigm: A Critique of the Adaptationist Programme," *Proceedings of the Royal Society* B 205 (1979): 581–98.

Gould, Stephen Jay and Elisabeth Vrba. "Exaptation: A Missing Term in the Science of Form," *Paleobiology* 8/1 (1982): 4–15.

Griffin, David. *Well Being: Its Meaning, Measurement and Moral Importance* (Oxford: Oxford University Press, 1986).

Grossman, S.P. *A Textbook of Physiological Psychology* (New York: John Wiley & Sons, 1973).

Hampton, Jean. *Hobbes and the Social Contract Tradition* (Cambridge: Cambridge University Press, 1988).

Hampton, Jean. "Two Faces of Contractarian Thought," in Peter Vallentyne (ed.), *Contractarianism and Rational Choice* (Cambridge: Cambridge University Press, 1991), 31–55.

Hampton, Jean. "Contract and Consent," in R. Goodin and P. Pettit (eds.), *A Companion to Contemporary Political Philosophy* (Oxford: Blackwell Publishers, 1993).

Hampton, Jean. *The Authority of Reason* (Cambridge: Cambridge University Press, 1998).

Hare, R.M. *Moral Thinking* (Oxford: Oxford University Press, 1981).

Harman, Gilbert. "Moral Relativism Defended," *Philosophical Review* 84 (1975): 3–22.

Harman, Gilbert. *The Nature of Morality* (New York: Oxford University Press, 1977).

Harsanyi, J. "Review of Gauthier's 'Morals by Agreement," *Economics and Philosophy* 3 (1987): 339–43.

Harsanyi, J. "Morality and the Theory of Rationality," in A. Sen and B. Williams (eds.), *Utilitarianism and Beyond* (Cambridge: Cambridge University Press, 1982), 44–60.

Hart, H.L.A. "Rawls on Liberty and Its Priority," *University of Chicago Law Review* 40 (1973): 535–55.

Haworth, Larry. *Autonomy: An Essay in Philosophical Psychology and Ethics* (New Haven: Yale University Press, 1986).

Heinrich, Joe, Wulf Albers, Robert Boyd, Gerd Gigerenzer, Kevin McCabe, Axel Ockenfels, and Peyton Young. "What is the Role of Culture in Bounded Rationality?" in Gerd Gigerenzer and R. Selten (eds.), *Bounded Rationality: The Adaptive Toolbox* (Cambridge, Mass.: MIT Press, 2002).

Hill, T., Jr. *Dignity and Practical Reason in Kant's Moral Theory* (Ithaca, N.Y.: Cornell University Press, 1992).

Hobbes, Thomas. *The Leviathan* (Buffalo: Prometheus Books, [1651] 1988).

Hobbes, Thomas. in Richard S. Peters (ed.), *Body, Man, and Citizen* (New York: Collier, 1962).

Homer, *The Odyssey* (Hammondsworth: Penguin, 1973).

Hubin, Donald. "Non-Tuism," *Canadian Journal of Philosophy* 21/4 (1991): 441–68.

Hull, David. "Individuality and Selection," *Annual Review of Ecology and Systematics* 11 (1980): 311–32.

Hull, David. "On Human Nature," in David Hull and Michael Ruse (eds.), *The Philosophy of Biology* (Oxford: Oxford University Press, 1998), 383–97.

Hull, David and Michael Ruse (eds.). *The Philosophy of Biology* (Oxford: Oxford University Press, 1998).

Hume, David. *A Treatise of Human Nature*, L.A. Selby-Bigge (ed.), 2nd edn revised by P.H. Nidditch (Oxford: Oxford University Press, [1888] 1978).

Hume, David. "Of the Original Contract," in Eugene Miller (ed.), *David Hume – Essays: Moral, Political, and Literary* (Indianapolis: LibertyClassics, 1987), 465–87.

Hume, David. "An Enquiry Concerning the Principles of Morals," in L.A. Selby-Bigge and P.H. Nidditch (eds.), *Enquiries Concerning Human Understanding and Concerning the Principles of Morals* (3rd edn) (Oxford: Clarendon Press, 1989).

Joyce, Richard. *The Myth of Morality* (Cambridge: Cambridge University Press, 2001).

Kahneman, D., P. Slovic, and A. Tversky (eds.). *Judgments under Uncertainty: Heuristics and Biases* (Cambridge: Cambridge University Press, 1979).

Kant, Immanuel. *Critique of Pure Reason*, Norman Kemp Smith (trans.) (New York: St. Martin's Press, 1929).

Kant, Immanuel. *Grounding for the Metaphysics of Morals*, James W. Ellington (trans.) (Indianapolis: Hackett Publishing Co., 1986).

Kant, Immanuel. "On the Common Saying: 'This May Be True in Theory, but it Does Not Apply in Practice'," in H. Reiss (ed.), *Kant: Political Writings* (Cambridge: Cambridge University Press, 1970).

Kavka, Gregory. "A Reconciliation Project," in D. Copp and D. Zimmerman (eds.), *Morality, Reason, and Truth* (Lanham, Md.: Rowman and Allenheld, 1984), 297–319.

Kavka, Gregory. *Hobbesian Moral and Political Theory* (Princeton: Princeton University Press, 1986).

Klosko, George. "Political Obligations and the Natural Duties of Justice," *Philosophy and Public Affairs* 23/3 (1994): 251–70.

Korsgaard, Christine. "Kant's Formula of Universal Law," *Pacific Philosophical Quarterly* 66 (1985): 24–47.

Korsgaard, Christine. *The Sources of Normativity* (Cambridge: Cambridge University Press, 1998).

Kraus, Jody. *The Limits of Hobbesian Contractarianism* (Cambridge: Cambridge University Press, 1993).

Kujundzic, Nebojsa. "On the Logic of Adjectives," *Dialogue: Canadian Philosophical Review* 40/Fall (2001): 803–9.

Kymlicka, Will. *Liberalism, Community, and Culture* (Oxford: Clarendon Press, 1989).

Kymlicka, Will. "The Social Contract Tradition," in Peter Singer (ed.), *A Companion to Ethics* (Oxford: Blackwell Publishers, 1991).

Laden, Anthony. "Games, Fairness, and Rawls's A Theory of Justice," *Philosophy and Public Affairs* 20 (1991): 189–222.

Latané, B. and J.M. Darley. *The Unresponsive Bystander: Why Does He Not Help?* (Englewood Cliffs: Prentice Hall, 1970).

Lerner, Melvin and D.J. Miller. "Just World Research and the Attribution Process: Looking Back and Ahead," *Psychological Bulletin* 85/5 (1978): 1030–51.

Lorenz, Konrad. *Studies in Animal and Human Behavior*, Robert Martin (trans.) (Cambridge: Cambridge University Press, 1970).

Lowry, Malcolm. *Under the Volcano* (Hammnondsworth: Penguin, [1947] 1980).

Luce, Duncan and Howard Raiffa. *Games and Decisions* (New York: Dover Publications, 1985).

MacIntosh, Duncan. "Persons and the Satisfaction of Preferences: Problems in the Rational Kinematics of Values," *The Journal of Philosophy* 40/4 (1993): 163–80.

MacIntyre, Alasdair. *After Virtue* (Notre Dame: University of Notre Dame Press, 1984).

MacIntyre, Alasdair. *Whose Justice? Which Rationality?* (Notre Dame: Notre Dame University Press, 1987).

Mackie, J.L. *Ethics: Inventing Right and Wrong* (Hammondsworth: Penguin, 1977).

Metz, Thaddeus. "The Reasonable and the Moral," *Social Theory and Practice* 28 (2002): 277–301.

Mill, John Stuart. *On Liberty* (Indianapolis: Hackett Publishing Co., [1859] 1978).

Mill, John Stuart. *Utilitarianism* (Buffalo, N.Y.: Prometheus Books, [1863] 1987).

Millgram, Elijah. "Williams' Argument Against External Reasons," *Noûs* 30/2 (1996): 197–220.

Moore, G.E. *Principia Ethica* (Cambridge: Cambridge University Press, [1903] 1986).

Moore, Margaret. *Foundations of Liberalism* (Oxford: Clarendon Press, 1993).

Morris, Christopher. "The Relation Between Self-Interest and Justice in Contractarian Ethics," *Social Philosophy and Policy* 5/2 (1988).

Morris, Christopher. "A Contractarian Account of Moral Justification," in Walter Sinnott-Armstrong and Mark Timmons (eds.), *Moral Knowledge? New Essays in Moral Epistemology* (Oxford: Oxford University Press, 1996), 215–43.

Morris, Christopher. *An Essay on the Modern State* (Cambridge: Cambridge University Press, 1998).

Morse, Jennifer Roback. "Who is Rational Economic Man?" in E. Frankel Paul, F. Miller, Jr., and J. Paul (eds.), *Self-Interest* (Cambridge: Cambridge University Press, 1997), 179–206.

Murray, Malcolm. "Unconsidered Preferences," *South African Journal of Philosophy* 17/4 (1998): 346–53.

Murray, Malcolm. "How to Blackmail a Contractarian," *Public Affairs Quarterly* 13/4 (1999): 347–63.

Murray, Malcolm. "Lewd Propositions and Contractarianism," in Kenneth Cust (ed.), *20th Century Values* (Value-Net, 2002), at http://value-net.org/Publications/20thCentury_Values/20thcentury_values.html.

Murray, Malcolm. "A Catalogue of Mistaken Interests: Reflections on the Desired and the Desirable," *International Journal of Philosophical Studies* 11/1 (2003): 1–23.

Murray, Malcolm. "Concerned Parties: When Lack of Consent is Irrelevant," *Public Affairs Quarterly* 18/2 (2004): 125–40.

Murray, Malcolm. "Why Contractarians are not Libertarians ... Evolutionarily Speaking," in Malcolm Murray (ed.), *Liberty, Games, and Contract: Jan Narveson and the Defence of Libertarianism* (Aldershot: Ashgate Press, 2007), 115–27.

Murray, Malcolm (ed.). *Liberty, Games, and Contract: Jan Narveson and the Defence of Libertarianism* (Aldershot: Ashgate Press, 2007).

Nagel, Thomas. *Mortal Questions* (Cambridge: Cambridge University Press, 1979).

Nagel, Thomas. *The View from Nowhere* (Oxford: Oxford University Press, 1986).

Narveson, Jan. "A Puzzle About Economic Justice in Rawls' Theory," *Social Theory and Practice* 4/1 (1976): 1–28.

Narveson, Jan. *The Libertarian Idea* (Philadelphia: Temple University Press, 1988).

Narveson, Jan. "The Anarchist's Case," in J.T. Sanders and J. Narveson (eds.), *For and Against the State* (Lanham, Md.: Rowman and Littlefield, 1996), 195–216.

Narveson, Jan. "Social Contract, Game Theory and Liberty: Responding to My Critics," in Malcolm Murray (ed.), *Liberty, Games, and Contract: Jan Narveson and the Defence of Libertarianism* (Aldershot: Ashgate Press, 2007), 217–40.

Nielsen, Kai. "Capitalism, Socialism, and Justice," in T. Regan and D. Van DeVeere (eds.), *And Justice for All* (Totowa, N.J.: Rowman & Allenheld, 1982), 264–86.

Nozick, Robert. *Anarchy, State, and Utopia* (New York: Basic Books, 1974).

Parfit, Derek. *Reasons and Persons* (Oxford: Oxford University Press, 1984).

Pineau, Lois. "Date Rape: A Feminist Analysis," *Law & Philosophy* 8 (1989): 217–43.

Pritchard, H.A. "Does Moral Philosophy Rest on a Mistake?" in H.A. Pritchard, *Moral Obligation* (Oxford: Clarendon Press, 1949).

Putnam, Hilary. *The Collapse of the Fact/Value Dichotomy and other Essays* (Cambridge, Mass.: Harvard University Press, 1992).

Railton, Peter. "Moral Realism: Prospects and Problems," in Walter Sinnott-Armstrong and Mark Timmons (eds.), *Moral Knowledge? New Readings in Moral Epistemology* (Oxford: Oxford University Press, 1996), 49–81.

Railton, Peter. "Moral Realism," *Philosophical Review* 95/2 (1986): 163–207.

Rawls, John. *A Theory of Justice* (Cambridge, Mass.: Harvard University Press, 1971).

Rawls, John. *Political Liberalism* (New York: Columbia University Press, 1993).

Raz, Joseph. *The Morality of Freedom* (Oxford: Oxford University Press, 1988).

Regan, T. and D. Van DeVeere (eds.). *And Justice for All* (Totowa, N.J.: Rowman & Allenheld, 1982).

Robinson, Daniel. *Praise and Blame: Moral Realism and Its Applications* (Princeton: Princeton University Press, 2002).

Rogers, Carl. *Client-Centered Therapy* (Boston: Houghton Mifflin, 1951).

Rorty, Richard. *Philosophy and the Mirror of Nature* (Princeton: Princeton University Press, 1979).

Rosenberg, Alexander. "Altruism: Theoretical Contexts," in E. Fox Keller and E. Lloyd (eds.), *Keywords in Evolutionary Biology* (Cambridge, Mass.: Harvard University Press, 1992), 19–28.

Ross, W.D. *The Right and the Good* (Oxford: Clarendon Press, 1930).

Sandel, Michael. *Liberalism and the Limits of Justice* (New York: Cambridge University Press, 1982).

Sayer-McCord, Geoffrey. "Coherentist Epistemology and Moral Theory," in Walter Sinnott-Armstrong and Mark Timmons (eds.), *Moral Knowledge? New Essays in Moral Epistemology* (Oxford: Oxford University Press, 1996), 137–89.

Scanlon, T.M. "Contractualism and Utilitarianism," in A. Sen and B. Williams (eds.), *Utilitarianism and Beyond* (Cambridge: Cambridge University Press, 1982), 103–28.

Scanlon, T.M. *What We Owe to Each Other* (Cambridge, Mass.: Belknap, Harvard University Press, 1998).

Schmidtz, David. "Rationality within Reason," *The Journal of Philosophy* 89/9 (1992): 445–66.

Schmidtz, David. "Choosing Ends," *Ethics* 104 (1994): 226–51.

Schmidtz, David. *Rational Choice and Human Agency* (Princeton: Princeton University Press, 1995).

Searle, John. "How to Derive an 'Ought' from an 'Is'," *Philosophical Review* 73 (1964): 43–58.

Selten, Reinhard. "What is Bounded Rationality?" in Gerd Gigerenzer and Reinhard Selten (eds.), *Bounded Rationality: The Adaptive Toolbox* (Cambridge, Mass.: MIT Press, 2002), 13–36.

Shafer-Landau, Russ. *Moral Realism: A Defence* (Oxford: Oxford University Press, 2003).

Shope, Robert. "The Conditional Fallacy in Contemporary Philosophy," *The Journal of Philosophy* 75/8 (1978).

Sidgwick, Henry. *The Methods of Ethics* (7th edn) (Cambridge: Hackett, [1874] 1981).

Singer, Peter. "Famine, Affluence, and Morality," *Philosophy and Public Affairs* 1 (1972): 229–43.

Sinnott-Armstrong, Walter, and Mark Timmons (eds.). *Moral Knowledge? New Readings in Moral Epistemology* (Oxford: Oxford University Press, 1996).

Skyrms, Brian. *Evolution of the Social Contract* (Cambridge: Cambridge University Press, 1996).

Skyrms, Brian. *The Stag Hunt and the Evolution of Social Structure* (Cambridge: Cambridge University Press, 2004).

Skyrms, Brian, and Jason Alexander. "Bargaining with Neighbors: Is Justice Contagious?" *Journal of Philosophy* 96/11 (1999): 588–98.

Smith, Holly. "Deriving Morality from Rationality," in P. Vallentyne (ed.), *Contractarianism and Rational Choice: Essays on David Gauthier's Morals By Agreement* (Cambridge: Cambridge University Press), 229–53.

Smith, John Maynard. "Science and Myth," in David Hull and Michael Ruse (eds.), *The Philosophy of Biology* (Oxford: Oxford University Press, 1998), 374–82.

Smith, Michael. *The Moral Problem* (Oxford: Blackwell, 1994).

Sober, Elliott. *Philosophy of Biology* (Boulder, Colo.: Westview Press, 1993).

Sober, Elliott. "What is Evolutionary Altruism?" *Canadian Journal of Philosophy* 14 (1988): 75–99.

Sober, Elliott and David Sloan Wilson. "A Critical Review of Philosophical Work on the Units of Selection Problem," *Philosophy of Science* 61 (1994): 534–55.

Sober, Elliott and David Sloan Wilson. *Unto Others: The Evolution and Psychology of Unselfish Behavior* (Cambridge, Mass.: Harvard University Press, 1998).

Spencer, Herbert. "Progress: Its Law and Cause," *Westminster Review* 9 (1857): 445–85.

Spencer, Herbert. *First Principles* (London: Williams and Norgate, 1862).

Sperba, Dan. *Explaining Culture: A Naturalistic Approach* (Oxford: Oxford University Press, 1996).

Stevenson, Charles. "The Emotive Meaning of Ethical Terms," *Mind* 46 (1937): 14–31.

Sumner, Wayne. "Something in Between," in Roger Crisp and Brad Hooker (eds.), *Well-Being and Morality: Essays in Honour of James Griffin* (Oxford University Press, 2000), 1–20.

Taylor, Charles. *Philosophy and the Human Sciences: Philosophical Papers 2* (Cambridge: Cambridge University Press, 1985).

Taylor, Charles. *Sources of the Self: The Making of Modern Identity* (Cambridge: Harvard University Press, 1989).

Taylor, Charles. *Human Agency and Language: Philosophical Papers 1* (Cambridge: Cambridge University Press, 1992).

Thompson, Judith. *The Realm of Rights* (Cambridge, Mass.: Harvard University Press, 1990).

Timmons, Mark. "Outline of a Contextualist Epistemology," in Walter Sinnott-Armstrong and Mark Timmons (eds.), *Moral Knowledge? New Readings in Moral Epistemology* (Oxford: Oxford University Press, 1996), 293–325.

Timmons, Mark. *Morality Without Foundations: A Defense of Ethical Contextualism* (Oxford: Oxford University Press, 1999).

Trivers, Robert. "The Evolution of Reciprocal Cooperation," *Quarterly Review of Biology* 46 (1971): 35–57.

Velleman, David. *Practical Reflection* (Princeton, Princeton University Press, 1989).

Viminitz, Paul. "A Defence of Terrorism," in Frederick R. Adams (ed.), *Ethical Issues for the Twenty-First Century* (Charlottesville: Philosophy Documentation Center Press, 2005).

Viminitz, Paul. "Getting the Baseline Right," in Malcolm Murray (ed.), *Liberty, Games and Contracts: Jan Narveson and the Defence of Libertarianism* (Aldershot: Ashgate Press, 2007), 129–43.

Walzer, Michael. *Interpretation and Social Criticism* (Cambridge, Mass.: Harvard University Press, 1987).

Watson, Gary. "Some Considerations in Favor of Contractualism," in Jules Coleman and Christopher Morris (eds.), *Rational Commitment of Social Justice: Essays for Gregory Kavka* (Cambridge: Cambridge University Press, 1998), 168–85.

Wells, H.G. *The Time Machine* (New York: Dover Publications, [1895] 1995).

Wiggins, David. *Needs, Values, Truth* (Oxford: Oxford University Press, 1998).

Williams, Bernard. "Internal and External Reasons," in Bernard Williams, *Moral Luck* (Cambridge: Cambridge University Press, 1981).

Williams, Bernard. *Ethics and the Limits of Philosophy* (Cambridge, Mass.: Harvard University Press, 1985).

Wilson, E.O. *Sociobiology: The New Synthesis* (Cambridge: Cambridge University Press, 1975).

Wolf, Susan. "Happiness and Meaning: Two Aspects of the Good Life," in E. Frankel Paul, F. Miller, Jr., and J. Paul (eds.), *Self-Interest* (Cambridge: Cambridge University Press, 1997), 207–25.

INDEX

PHILOSOPHICAL STUDIES SERIES

1. Jay F. Rosenberg: *Linguistic Representation.* 1974 ISBN 90-277-0533-X
2. Wilfrid Sellars: *Essays in Philosophy and Its History.* 1974 ISBN 90-277-0526-7
3. Dickinson S. Miller: *Philosophical Analysis and Human Welfare.* Selected Essays and Chapters from Six Decades. Edited with an Introduction by Lloyd D. Easton. 1975 ISBN 90-277-0566-6
4. Keith Lehrer (ed.): *Analysis and Metaphysics.* Essays in Honor of R. M Chisholm. 1975 ISBN 90-277-0571-2
5. Carl Ginet: *Knowledge, Perception, and Memory.* 1975 ISBN 90-277-0574-7
6. Peter H. Hare and Edward H. Maddern: *Causing. Perceiving and Believing.* An Examination of the Philosophy of C.J. Ducasse. 1975 ISBN 90-277-0563-1
7. Hector-Neri Castañeda: *Thinking and Doing.* The Philosophical Foundations of Institutions. 1975 ISBN 90-277-0610-7
8. John L. Pollock: *Subjunctive Reasoning.* 1976 ISBN 90-277-0701-4
9. Bruce Aune: *Reason and Action.* 1977 ISBN 90-277-0805-3
10. George Schlesinger: *Religion and Scientific Method.* 1977 ISBN 90-277-0815-0
11. Yirmiahu Yovel (ed.): *Philosophy of History and Action.* Papers presented at the First Jerusalem Philosophical Encounter (December 1974). 1978 ISBN 90-277-0890-8
12. Joseph C. Pitt (ed.): *The Philosophy of Wilfrid Sellars: Queries and Extensions.* 1978 ISBN 90-277-0903-3
13. Alvin I. Goldman and Jaegwon Kim (eds.): *Values and Morals.* Essays in Honor of William Frankena. Charles Stevenson, and Richard Brandt. 1978 ISBN 90-277-0914-9
14. Michael J. Loux: *Substance and Attribute.* A Study in Ontology. 1978 ISBN 90-277-0926-2
15. Ernest Sosa (ed.): *The Philosophy of Nicholas Rescher.* Discussion and Replies. 1979 ISBN 90-277-0962-9
16. Jeffrie G. Murphy: *Retribution, Justice, and Therapy.* Essays in the Philosophy of Law. 1979 ISBN 90-277-0998-X
17. George S. Pappas (ed.): *Justification and Knowledge.* New Studies in Epistemology. 1979 ISBN 90-277-1023-6
18. James W. Cornman: *Skepticism, Justification, and Explanation.* With a Bibliographic Essay by Walter N. Gregory. 1980 ISBN 90-277-1041-4
19. Peter van Inwagen (ed.): *Time and Cause.* Essays presented to Richard Taylor. 1980 ISBN 90-277-1048-1
20. Donald Nute: *Topics in Conditional Logic.* 1980 ISBN 90-277-1049-X
21. Risto Hilpinen (ed.): *Rationality in Science.* Studies in the Foundations of Science and Ethics. 1980 ISBN 90-277-1112-7
22. George Dicker: *Perceptual Knowledge.* An Analytical and Historical Study. 1980 ISBN 90-277-1130-5

PHILOSOPHICAL STUDIES SERIES

23. Jay F. Rosenberg: *One World and Our Knowledge of It*. The Problematic of Realism in Post-Kantian Perspective. 1980 ISBN 90-277-1136-4
24. Keith Lehrer and Carl Wagner: *Rational Consensus in Science and Society*. A Philosophical and Mathematical Study. 1981 ISBN 90-277-1306-5
25. David O'Connor: *The Metaphysics of G. E. Moore*. 1982 ISBN 90-277-1352-9
26. John D. Hodson: *The Ethics of Legal Coercion*. 1983 ISBN 90-277-1494-0
27. Robert J. Richman: *God, Free Will, and Morality*. Prolegomena to a Theory of Practical Reasoning. 1983 ISBN 90-277-1548-3
28. Terence Penelhum: *God and Skepticism*. A Study in Skepticism and Fideism. 1983 ISBN 90-277-1550-5
29. James Bogen and James E. McGuire (eds.): *How Things Are*. Studies in Predication and the History of Philosophy of Science. 1985 ISBN 90-277-1583-1
30. Clement Dore: *Theism*. 1984 ISBN 90-277-1683-8
31. Thomas L. Carson: *The Status of Morality*. 1984 ISBN 90-277-1619-9
32. Michael J. White: *Agency and Integrality*. Philosophical Themes in the Ancient Discussions of Determinism and Responsibility. 1985 ISBN 90-277-1968-3
33. Donald F. Gustafson: *Intention and Agency*. 1986 ISBN 90-277-2009-6
34. Paul K. Moser: *Empirical Justification*. 1985 ISBN 90-277-2041-X
35. Fred Feldman: *Doing the Best We Can*. An Essay in Informal Deontic Logic. 1986 ISBN 90-277-2164-5
36. G. W. Fitch: *Naming and Believing*. 1987 ISBN 90-277-2349-4
37. Terry Penner: *The Ascent from Nominalism*. Some Existence Arguments in Plato's Middle Dialogues. 1987 ISBN 90-277-2427-X
38. Robert G. Meyers: *The Likelihood of Knowledge*. 1988 ISBN 90-277-2671-X
39. David F. Austin (ed.): *Philosophical Analysis*. A Defense by Example. 1988 ISBN 90-277-2674-4
40. Stuart Silvers (ed.): *Rerepresentation*. Essays in the Philosophy of Mental Representation. 1988 ISBN 0-7923-0045-9
41. Michael P. Levine: *Hume and the Problem of Miracles*. A Solution. 1989 ISBN 0-7923-0043-2
42. Melvin Dalgarno and Eric Matthews (eds.): *The Philosophy of Thomas Reid*. 1989 ISBN 0-7923-0190-0
43. Kenneth R. Westphal: *Hegel's Epistemological Realism*. A Study of the Aim and Method of Hegel's *Phenomenology of Spirit*. 1989 ISBN 0-7923-0193-5
44. John W. Bender (ed.): *The Current State of the Coherence Theory*. Critical Essays on the Epistemic Theories of Keith Lehrer and Laurence BonJour, with Replies. 1989 ISBN 0-7923-0220-6
45. Roger D. Gallie: *Thomas Reid and 'The Way of Ideas'*. 1989 ISBN 0-7923-0390-3

PHILOSOPHICAL STUDIES SERIES

46. J-C. Smith (ed.): *Historical Foundations of Cognitive Science*. 1990
ISBN 0-7923-0451-9
47. John Heil (ed.): *Cause, Mind, and Reality*. Essays Honoring C. B. Martin. 1989
ISBN 0-7923-0462-4
48. Michael D. Roth and Glenn Ross (eds.): *Doubting*. Contemporary Perspectives on Skepticism. 1990 ISBN 0-7923-0576-0
49. Rod Bertolet: *What is Said*. A Theory of Indirect Speech Reports. 1990
ISBN 0-7923-0792-5
50. Bruce Russell (ed.): *Freedom, Rights and Pornography*. A Collection of Papers by Fred R. Berger. 1991 ISBN 0-7923-1034-9
51. Kevin Mulligan (ed.): *Language, Truth and Ontology*. 1992 ISBN 0-7923-1509-X
52. Jesús Ezquerro and Jesús M. Larrazabal (eds.): *Cognition, Semantics and Philosophy*. Proceedings of the First International Colloquium on Cognitive Science. 1992
ISBN 0-7923-1538-3
53. O.H. Green: *The Emotions*. A Philosophical Theory. 1992 ISBN 0-7923-1549-9
54. Jeffrie G. Murphy: *Retribution Reconsidered*. More Essay in the Philosophy of Law. 1992 ISBN 0-7923-1815-3
55. Phillip Montague: *In the Interests of Others*. An Essay in Moral Philosophy. 1992
ISBN 0-7923-1856-0
56. Jacques-Paul Dubucs (ed.): *Philosophy of Probability*. 1993 ISBN 0-7923-2385-8
57. Gary S. Rosenkrantz: *Haecceity*. An Ontological Essay. 1993 ISBN 0-7923-2438-2
58. Charles Landesman: *The Eye and the Mind*. Reflections on Perception and the Problem of Knowledge. 1994 ISBN 0-7923-2586-9
59. Paul Weingartner (ed.): *Scientific and Religious Belief*. 1994 ISBN 0-7923-2595-8
60. Michaelis Michael and John O'Leary-Hawthorne (eds.): *Philosophy in Mind*. The Place of Philosophy in the Study of Mind. 1994 ISBN 0-7923-3143-5
61. William H. Shaw: *Moore on Right and Wrong*. The Normative Ethics of G.E. Moore. 1995 ISBN 0-7923-3223-7
62. T.A. Blackson: *Inquiry, Forms, and Substances*. A Study in Plato's Metaphysics and Epistemology. 1995 ISBN 0-7923-3275-X
63. Debra Nails: *Agora, Academy, and the Conduct of Philosophy*. 1995
ISBN 0-7923-3543-0
64. Warren Shibles: *Emotion in Aesthetics*. 1995 ISBN 0-7923-3618-6
65. John Biro and Petr Kotatko (eds.): *Frege: Sense and Reference One Hundred Years Later*. 1995 ISBN 0-7923-3795-6
66. Mary Gore Forrester: *Persons, Animals, and Fetuses*. An Essay in Practical Ethics. 1996 ISBN 0-7923-3918-5
67. K. Lehrer, B.J. Lum, B.A. Slichta and N.D. Smith (eds.): *Knowledge, Teaching and Wisdom*. 1996 ISBN 0-7923-3980-0

PHILOSOPHICAL STUDIES SERIES

68. Herbert Granger: *Aristotle's Idea of the Soul*. 1996 ISBN 0-7923-4033-7
69. Andy Clark, Jesús Ezquerro and Jesús M. Larrazabal (eds.): *Philosophy and Cognitive Science: Categories, Consciousness, and Reasoning*. Proceedings of the Second International Colloquium on Cognitive Science. 1996 ISBN 0-7923-4068-X
70. J. Mendola: *Human Thought*. 1997 ISBN 0-7923-4401-4
71. J. Wright: *Realism and Explanatory Priority*. 1997 ISBN 0-7923-4484-7
72. X. Arrazola, K. Korta and F.J. Pelletier (eds.): *Discourse, Interaction and Communication*. Proceedings of the Fourth International Colloquium on Cognitive Science. 1998 ISBN 0-7923-4952-0
73. E. Morscher, O. Neumaier and P. Simons (eds.): *Applied Ethics in a Troubled World*. 1998 ISBN 0-7923-4965-2
74. R.O. Savage: *Real Alternatives, Leibniz's Metaphysics of Choice*. 1998 ISBN 0-7923-5057-X
75. Q. Gibson: *The Existence Principle*. 1998 ISBN 0-7923-5188-6
76. F. Orilia and W.J. Rapaport (eds.): *Thought, Language, and Ontology*. 1998 ISBN 0-7923-5197-5
77. J. Bransen and S.E. Cuypers (eds.): *Human Action, Deliberation and Causation*. 1998 ISBN 0-7923-5204-1
78. R.D. Gallie: *Thomas Reid: Ethics, Aesthetics and the Anatomy of the Self*. 1998 ISBN 0-7923-5241-6
79. K. Korta, E. Sosa and X. Arrazola (eds.): *Cognition, Agency and Rationality*. Proceedings of the Fifth International Colloquium on Cognitive Science. 1999 ISBN 0-7923-5973-9
80. M. Paul: *Success in Referential Communication*. 1999 ISBN 0-7923-5974-7
81. E. Fischer: *Linguistic Creativity*. Exercises in 'Philosophical Therapy'. 2000 ISBN 0-7923-6124-5
82. R. Tuomela: *Cooperation*. A Philosophical Study. 2000 ISBN 0-7923-6201-2
83. P. Engel (ed.): *Believing and Accepting*. 2000 ISBN 0-7923-6238-1
84. W.L. Craig: *Time and the Metaphysics of Relativity*. 2000 ISBN 0-7923-6668-9
85. D.A. Habibi: *John Stuart Mill and the Ethic of Human Growth*. 2001 ISBN 0-7923-6854-1
86. M. Slors: *The Diachronic Mind*. An Essay on Personal Identity. Psychological Continuity and the Mind-Body Problem. 2001 ISBN 0-7923-6978-5
87. L.N. Oaklander (ed.): *The Importance of Time*. Proceedings of the Philosophy of Time Society. 1995–2000. 2001 ISBN 1-4020-0062-6
88. M. Watkins: *Rediscovering Colors*. A Study in Pollyanna Realism. 2002 ISBN 1-4020-0737-X
89. W.F. Vallicella: *A Paradigm Theory of Existence*. Onto–Theology Vindicated. 2002 ISBN 1-4020-0887-2

PHILOSOPHICAL STUDIES SERIES

90. M. Hulswit: *From Cause to Causation*. A Peircean Perspective. 2002
ISBN 1-4020-0976-3; Pb 1-4020-0977-1
91. D. Jacquette (ed.): *Philosophy, Psychology, and Psychologism*. Critical and Historical Readings on the Psychological Turn in Philosophy. 2003 ISBN 1-4020-1337-X
92. G. Preyer, G. Peter and M. Ulkan (eds.): *Concepts of Meaning*. Framing an Integrated Theory of Linguistic Behavior. 2003 ISBN 1-4020-1329-9
93. W. de Muijnek: *Dependencies, Connections, and Other Relations*. A Theory of Mental Causation. 2003 ISBN 1-4020-1391-4
94. N. Milkov: *A Hundred Years of English Philosophy*. 2003 ISBN 1-4020-1432-5
95. E.J. Olsson (ed.): *The Epistomology of Keith Lehrer*. 2003 ISBN 1-4020-1605-0
96. D.S. Clarke: *Sign Levels*. Language and Its Evolutionary Antecedents. 2003
ISBN 1-4020-1650-6
97. A. Meirav: *Wholes, Sums and Unities*. 2003 ISBN 1-4020-1660-3
98. C.H. Conn: *Locke on Esence and Identity*. 2003 ISBN 1-4020-1670-0
99. J.M. Larrazabal and L.A. Pérez Miranda (eds.): *Language, Knowledge, and Representation*. Proceedings of the Sixth International Colloquium on Cognitive Science (ICCS-99). 2004 ISBN 1-4020-2057-0
100. P. Ziff: *Moralities*. A Diachronic Evolution Approach. 2004 ISBN 1-4020-1891-6
101. J.A. Corlett: Terrorism: *A Philosophical Analysis*. 2003
ISBN 1-4020-1694-8: Pb 1-4020-1695-6
102. K. Korta and J.M. Larrazabal (eds): *Truth, Rationality, Cognition, and Music*. Proceedings of the Seventh International Colloquium on Cognitive Science. 2004
ISBN 1-4020-1912-2
103. T.M. Crisp, M. Davidson and D. Vander Laan (eds.): *Knowledge and Reality*. 2006
ISBN 1-4020-4732-0
104. S. Boër: *Thought-Contents*: On the Ontology of Belief and the Semantics of Belief Attribution. 2006 ISBN 1-4020-5084-4
105. A. Voltolini: *How Ficta Follow Fiction*. 2006 ISBN 1-4020-5146-8
106. K.L. Miller (ed.): *Issues in Theoretical Diversity*. Persistence, Composition, and Time. 2006 ISBN 1-4020-5255-3
107. M. Kieran (eds.): *Knowing Art*. Essays in Aesthetics and Epistemology. 2006
ISBN 1-4020-5264-2
108. M. Murray (ed.): *The Moral Wager*. Evolution and Contract. 2007
ISBN 978-1-4020-5854-7

Printed in the United States
104046LV00003B/1/A